The Myth of Resource Efficiency
The Jevons Paradox

The Myth of Resource Efficiency

The Jevons Paradox

John M. Polimeni, Kozo Mayumi,
Mario Giampietro and Blake Alcott

from Routledge

Originally published in 2008 as *The Jevons Paradox and the Myth of Resource Efficiency Improvements*

First published by Earthscan in the UK and USA in 2009

For a full list of publications please contact:

Earthscan
2 Park Square, Milton Park, Abingdon, Oxon OX14 4RN
711 Third Avenue, New York, NY 10017

Earthscan is an imprint of the Taylor & Francis Group, an informa business

Notices

Practitioners and researchers must always rely on their own experience and knowledge in evaluating and using any information, methods, compounds, or experiments described herein. In using such information or methods they should be mindful of their own safety and the safety of others, including parties for whom they have a professional responsibility.

Product or corporate names may be trademarks or registered trademarks, and are used only for identification and explanation without intent to infringe.

ISBN 978-1-84407-813-4 (pbk)

Typeset by Domex e-Data, India
Cover design by Susanne Harris

A catalogue record for this book is available from the British Library
Library of Congress Cataloging-in-Publication Data has been applied for

Contents

List of Figures and Tables

FIGURES

TABLES

Foreword

Joseph A. Tainter

It is provided in the essence of things that from any fruition of success, no matter what, shall come forth something to make a greater struggle necessary.

Walt Whitman, *Song of the Open Road*

In January 2002 I was returning to the United States from fieldwork in the Sahel of Mali. My itinerary to and from Mali goes through Paris, where usually I take a layover. On this occasion I arranged to meet friends for dinner, at which we were joined by a Swedish geographer. The conversation turned to various topics, including the platform of the Green Party in upcoming elections. Since we were discussing environmental issues, our Swedish colleague told us about a study he had recently done. It was a project of survey research, in which Swedes had been asked the question, 'If you were to eat less meat in your daily diet, what would you do with the money this saves?' It turns out that if Swedes ate less meat, they would like to use the money to travel more. Travel, of course, carries environmental costs, just as does eating meat. Reducing consumption of meat might not reduce environmental damage and certainly wouldn't eliminate it, a somewhat counter-intuitive outcome. But that is the nature of the Jevons Paradox. An action taken to conserve resources reduces the cost of daily life to such an extent that entirely different kinds of environmental damage become affordable. William Stanley Jevons would have predicted it.

In his 1865 work *The Coal Question*, William Stanley Jevons (1835–1882) expressed the concern that Britain would lose its economic dynamism and pre-eminence in the world due to an inevitable depletion of its reserves of easily mined coal. Of course he did not foresee the dominance of petroleum, even denying its likelihood, and so the central worry of the book turned out to be misplaced. But *The Coal Question* contains a gem that enshrines the book as among the most significant works of resource economics. That gem is known today as the Jevons Paradox. It cannot be expressed better than in Jevons's own Victorian prose:

It is wholly a confusion of ideas to suppose that the economical use of fuel is equivalent to a diminished consumption. The very contrary is the truth. (Jevons, 1866, p123)

> *As a rule, new modes of economy will lead to an increase of consumption ...* (Jevons, 1866, p123)

> *Now, if the quantity of coal used in a blast-furnace, for instance, be diminished in comparison with the yield, the profits of the trade will increase, new capital will be attracted, the price of pig-iron will fall, but the demand for it increase; and eventually the greater number of furnaces will more than make up for the diminished consumption of each.* (Jevons, 1866, p124–125)

In short, as technological improvements increase the efficiency with which a resource is used, total consumption of that resource may increase rather than decrease. This paradox has implications of the highest importance for the energy future of industrialized nations. It suggests that efficiency, conservation and technological improvement, the very things urged by those concerned for future energy supplies, may actually worsen our energy prospects.

The present book is one of the most extraordinary works on the Jevons Paradox. The authors are known for their innovative and eclectic research. The topics covered here are diverse, as are the approaches of the individual chapters. Blake Alcott in Chapter 2 sets the historical scene, discussing Jevons's work in the context of the founders of economics in the 18th and early 19th centuries. Mario Giampietro and Kozo Mayumi in Chapter 3 continue their explorations of epistemology and societal energy metabolism from a thermodynamic perspective. They discuss the important trade-off between efficiency and adaptability, referring to this as a yin–yang tension. In the fourth chapter, John Polimeni conducts a technical analysis to determine whether the Jevons Paradox has been in effect in various countries and regions of the world. The individual chapters are technical, and are valuable for this. They investigate in a rigorous manner the question of whether industrial nations can expect to continue in their present mode based on the hope and expectation of increasing efficiency in energy use.

The Jevons Paradox questions the pervasive assumption – common in colloquial discourse and even in many academic discussions – that sustainability emerges as a passive consequence of consuming less. This assumption comes in two versions. The pessimistic version suggests that it is necessary for people voluntarily to reduce their resource consumption in order to become more sustainable. Examples might include taking shorter or colder showers, using public transportation, drinking tap water rather than bottled, or eating less meat. This is sometimes known as the sackcloth-and-ashes approach to sustainability. The optimistic version, preferred by many economists and most politicians, is that a future of technological innovations and the shift to a service-and-information economy will reduce our consumption of resources to such an extent that we will become sustainable without requiring people to sacrifice the things that they enjoy. In this view of the future, technical improvements will allow us to produce more gross domestic product per unit of resource consumption than at present, and thereby maintain our way of life. This is exactly the assumption that Jevons showed to be false in the third quotation above. In his day, the assumption of a technical solution involved blast furnaces, coal and pig iron. Today the assumption involves energy and our way of life in the broadest possible sense. The Jevons Paradox is based on a foundation principle of economics: any time one reduces the cost of consuming a valued resource, people will respond by consuming more of it. Or, as suggested in the opening paragraph of this essay, people will consume more of something else, perhaps resulting in no net savings or even greater overall consumption. As the noted journalist Eric Sevareid once said, 'The chief cause of problems is solutions.'

As Blake Alcott shows in this volume, the Jevons Paradox is connected to the work of other distinguished writers in the history of economics. Kenneth Boulding, for example, once developed three theorems from the work of Thomas Robert Malthus, which he presented in a foreword to Malthus's *Population: The First Essay*. Boulding labelled his first theorem the Dismal Theorem:

> *If the only ultimate check on the growth of population is misery, then the population will grow until it is miserable enough to stop its growth.* (Boulding, 1959, pvii)

Theorem two is the Utterly Dismal Theorem:

> *Any technical* improvement *can only relieve misery for a while, for as long as misery is the only check on population, the improvement will enable population to grow, and will soon enable* more *people to live in misery than before. The final result of improvements, therefore, is to increase the equilibrium population, which is to increase the sum total of human misery.* (Boulding, 1959, pvii; emphasis original)

Boulding's third theorem is called the Moderately Cheerful Form of the Dismal Theorem:

> *If something else, other than misery and starvation, can be found which will keep a prosperous population in check, the population does not have to grow until it is miserable and starves, and can be stably prosperous.* (Boulding, 1959, pxi)

Boulding observed that how to implement the Moderately Cheerful Theorem 'is a problem which has so far produced no wholly satisfactory solution' (1959, pxi).

One recognizes, of course, that these theorems are not confined to population. The Utterly Dismal Theorem in particular is quite consistent with the Jevons Paradox, and seems indeed to be a limited restatement of it. Boulding confined the theorem to technical improvements and population, but as Jevons's analysis implies, the same principle applies to efficiency improvements in any costly thing that people acquire, whether children, automobiles or steaks. Reduce the cost of raising children, Jevons and Boulding would suggest, and people will raise more of them.

The Jevons Paradox has influenced world history, of which the Roman Empire provides an illustration. Early in its history, when it was a small city-state, Rome fought wars for survival against its immediate neighbours. Over time Rome was successful, defeating and subjugating these challengers. Early on the Romans adopted a clever strategy: incorporating the wealth and manpower of defeated rivals into Rome's war machine. In return, former rivals within Italy were given carefully graded rights in Rome's legal and political systems. Each time Rome defeated a rival, it emerged not only safer but stronger as well. When the time came in the third century BC for Rome to expand out of Italy, it had much of the resources of Italy at its command. This included great supplies of manpower. As Rome's empire expanded to the whole of the Mediterranean Basin and northwestern Europe, it continued most of this strategy, turning the resources of conquered nations to its own use. In 167 BC, for example, the Romans captured the Macedonian treasury, and promptly eliminated taxation of themselves. When Pergamon was annexed in 130 BC the state budget was doubled. After conquering Syria in 63 BC, Pompey raised the budget another 70 per cent. Julius Caesar relieved the Gauls of so much gold that its value in Rome fell 36 per cent (Tainter, 1988). In the terms of the Jevons Paradox, Rome's strategy resulted in a great reduction of the cost of conquest. The conquered nations underwrote the cost of Rome's further expansion. Finding conquest so economical, Rome responded by conquering more. A new mode of economy in conquest, Jevons would have observed, led not to contentment, but to an increase in the conquest of rival states. 'It is the very economy of [conquest],' Jevons might have written had he addressed Roman history, 'that leads to its extensive [employment]' (1866, p124).

Beyond such weighty matters as population, resources, and the fates of nations, the Jevons Paradox can be found in operation in many matters of daily life, both great and small. Since the authors of this volume have analysed technical aspects of the Jevons Paradox, I will take the opportunity to explore some examples from *la vie quotidienne*. I am presently at a keyboard, so word processing comes to mind. Early in my career, when professional writing had to be done on a typewriter, it was a costly endeavour to produce a new version of even a paper of ordinary length, let alone a book. The cost was accounted in time, labour, sore muscles of the hands and shoulders, and mental fatigue. When I acquired my first personal computer with a word processor in 1983, I thought naively that it would save me a great deal of work. Many other early adopters thought similarly. It was widely stated at the time that as a society we would now save great quantities of paper – the paperless office, as it came to be called.

Just the opposite has proved to be the case. Word processing has so reduced the cost of producing a single draft of a text that I now edit and generate six to eight drafts of everything I write for publication. In typewriter days I would usually produce only two drafts. While I have not kept a log, I strongly suspect that the amounts of time and labour that I invest producing a text has increased with the availability of the time – and the labour-saving word processor. As for paper, most drafts get printed, so that I consume much more paper than ever. When was the last time anyone predicted the paperless office? New modes of economy such as word processing, as Jevons noted, lead to an increase in consumption and even an increase in work.

The Jevons Paradox affects law enforcement officers who need to subdue violent suspects. A police officer who shoots a suspect with a gun pays a personal cost that is potentially very high. The officer is typically suspended from duty for a few days while a review board investigates the shooting. This period of review no doubt exacts a high emotional cost. The officer could be found to have discharged his or her firearm improperly. In that case, the officer might be dismissed, sued by the suspect or the suspect's family, or even prosecuted and imprisoned. Over the past few years, many police forces in the United States have equipped their officers with a device known as the Taser. The gun and the Taser are both high-energy devices capable of subduing suspects. The Taser imparts an electric shock of 50,000 volts, momentarily disabling a suspect, but usually not fatally. It can even be fired from several metres away. The Taser was initially presented as a humane device, which would allow police to subdue a violent offender without having to use a gun. The overlooked advantage is that the Taser has reduced the personal cost and risk to police officers of employing a high-energy weapon to overpower an offender. Unless the Tasered suspect dies, or the incident is recorded on video (both of which happen rarely), no board reviews ordinary use of the device. Since officers now face less personal cost if they use a high-energy weapon, they will be inclined to use such a weapon more often. Tasers are found at this writing in 11,500 US police forces. Predictably, there are now increasing numbers of complaints that police use Tasers too often (*USA Today*, 2007). Whether or not Tasers are used inappropriately, they are clearly used often enough to generate controversy. Jevons would not have found this surprising. Had he known of such a thing in his day, perhaps he would have written something like 'It is wholly a confusion of ideas to suppose that [reducing the personal cost of using a high-energy weapon] will lead to a diminished [usage of high-energy weapons]. The very contrary is the truth' (Jevons, 1866, p123).

My home community of Corrales, New Mexico has, like any responsible municipality, a board that oversees matters of planning and zoning. Corrales is a small community, but populated by intelligent, creative people, including many who work at the University of New Mexico and Sandia National Laboratories. They are an entrepreneurial population, and the village is home to many small businesses run from peoples' homes. The Planning and Zoning Board is responsible for approving home occupation permits. One evening when I attended a meeting of this board, I witnessed a proposal from a gentleman who wanted to operate a business from his home placing and servicing soft drink vending machines. It was to be a source of retirement income. His specialty was to place the machines in small offices where only a few people work. How, one might wonder, could one profit from placing these machines in small offices? The answer is that technical innovations in vending machines have reduced their energy consumption. Newer machines meet the Energy Star requirements of the United States Environmental Protection Agency. With reduced energy consumption, the machines can now be operated at a profit even in places where only a handful of people per day might purchase a soft drink. Newer machines even come with motion detectors, to turn on the front panel lighting when a potential customer approaches. These machines will also monitor the ambient temperature and 'learn' customers' habits. One brand claims to save 46 per cent of operating costs per year (www.vendingmiserstore.com). What is the outcome of all of this saving of energy? These machines are now to be found in small offices and other places where previously they would have been uneconomical. There are many more small offices than large ones, so that the population of vending machines is larger than ever. Is the net effect to save energy or to use more? Consider how Jevons might have phrased it: 'The profits of the trade will increase, new capital will be attracted, the price of [vending] will fall, but the demand for it increase; and eventually the greater number of [vending machines] will more than make up for the diminished consumption of each' (Jevons, 1866, pp124–125).

The United States has an antiquated system of air traffic control, with computer technology and displays dating from the days of vacuum tubes. A new system has been in preparation for years. It will, of course, feature updated electronic wizardry, but that is just the start. Currently, large commercial planes fly a 'post-to-post' system. That is, they fly a straight line to a certain point, governed by ground radar, then alter their course slightly to the next point, and so on across the land. The effect is that planes must fly a slightly zigzag course, which increases their time in transit and the fuel that they consume. Newer technology will enable planes to fly with the aid of global positioning system satellites. This will eliminate the need for a zigzag course. Planes will be able to fly straight to their destinations (that is, 'straight' within the constraints of the curvature of the earth), reducing time in the air and making air traffic more efficient overall. Pilots on approach will not need to maintain the large distance between planes that they do now. The system is called Automatic Dependent Surveillance-Broadcast (ADS-B). An early implementation at United Parcel Service's hub airport at Louisville, Kentucky, shows shorter taxi times, and steep cuts in emissions and noise as fewer planes must linger in the air awaiting a chance to land. UPS expects to save 900,000 gallons of fuel a year on 117 planes (Doyle and Gillies, 2007). In the normal way that such technical developments are viewed, this will be seen as a great improvement. We will enjoy more efficient air transport and safer air travel, and we will save time and fuel. But will this save time and fuel in the long run? Consider again Jevons's insight: 'As a rule, new modes of economy will lead to an increase of consumption' (1866, p123). Whatever savings will be brought by ADS-B will no doubt encourage even greater use of air travel and air freight. In the long run, the time spent in air travel will increase, as will the fuel that it consumes. As Boulding might have written, 'As long as misery is the only check on [air travel], the improvement will enable [air travel] to grow, and will soon merely enable *more* people to [travel] in misery than before' (1959, pvii).

What is to be done about the Jevons Paradox? It is a common human tendency to think locally and short-term (Tainter, 2007). In our history as a species, there was never selective pressure to think in terms of broader scales of space and time. Since humans did not evolve to think broadly, most of us don't. This suggests that people will not forgo currently affordable consumption on the basis of abstract projections about future resource supplies. Thus the Jevons Paradox cannot be circumvented through voluntary restraint or any other *laissez-faire* approach. Giampietro and Mayumi suggest that taxes could make up for any savings introduced by efficiency improvements, thereby avoiding the paradox. In the United States, at least, this approach is politically infeasible, but the general point is sound: The key to avoiding the Jevons Paradox is to adopt the principle that neither efficiency improvements, nor any other approach to reducing resource use (including voluntary conservation), can be allowed to reduce the cost of consumption. This is one way to implement the Moderately Cheerful Form of the Dismal Theorem. It is a principle that can be shown to work, illustrated again through my own experience.

In 1992, the desert city of Albuquerque, New Mexico, discovered that the aquifer from which it was drawing its water was much shallower than had been thought. Albuquerque had been planning to use this aquifer for the city's future growth. Now it was clear that those plans would need to be changed. To their credit, the city's officials acted immediately, implementing a number of conservation measures (including fines for excessive use) and arranging to replace groundwater with surface water. The programme had early success: people reduced their water consumption, so much so that the city's water utility suddenly found itself with insufficient income at a time when new investments were required. It was necessary to raise water rates. People naturally complained: they had done their duty by conserving water but realized no monetary reward for doing so. Instead they paid more to use less. The fortuitous part of this dilemma is that higher rates gave people a continuing incentive to conserve. And conserve they have continued to do. In 2005, Albuquerque recorded its lowest water consumption since 1985, even though its population has grown by 33 per cent in that time (*US Water News*, 2005). The city continues to encourage people to conserve, and the consistency of this message has no doubt helped. But the increase in water rates and the fines have helped too, circumventing the Jevons Paradox.

John Polimeni goes to great effort in his chapter to investigate nations and regions where the Jevons Paradox is in effect. I applaud his research, which is necessary to satisfy technical specialists. Yet the brief discussion here suggests that we might reverse the question and ask: Where is the Jevons Paradox *not* in effect?

REFERENCES

Boulding, K. E. (1959) 'Foreword', in Thomas Robert Malthus (author) *Population: The First Essay*, University of Michigan Press, Ann Arbor, MI, ppv–xii

Doyle, T. and Gillies, A. T. (2007) 'Smarter skies', *Forbes*, vol 179, no 4, available at http://members.forbes.com/forbes/2007/0226/055.html

Jevons, W. S. (1866) *The Coal Question: An Inquiry Concerning the Progress of the Nation and the Probable Exhaustion of our Coal-Mines*, 2nd edition, Macmillan, London

Tainter, J. A. (1988) *The Collapse of Complex Societies*, Cambridge University Press, Cambridge.

Tainter, J. A. (2007) 'Scale and dependency in world systems: Local societies in convergent evolution', in Alf Hornborg, J. R. McNeill and Joan Martinez-Alier (eds) *Rethinking Environmental History: World System History and Global Environmental Change*, AltaMira Press, Lanham, MD, pp361–377

USA Today (2007) 'Taser incidents renew debate over usage', article by John Curran on *USA Today* website, available at www.usatoday.com/news/nation/2007-08-23-1769869495_x.htm (accessed 17 October 2007)

US Water News (2005) 'Albuquerque water use at lowest since 1985', *US Water News* online edition, January 2005, available at www.uswaternews.com/archives/arcconserv/5albuwate1.html (accessed 17 October 2007)

1

Introduction

Energy policy is one of the most important issues facing the world today. This can be easily explained by looking at three dramatic changes taking place on our planet: first, the explosion of human population – in the last century, world population has more than tripled from approximately 1.7 billion people in 1900 to more than 6 billion people in 2000.[1] Even more impressive has been the increase in the pace of population growth. World population has grown from 3.5 billion people at the beginning of the 1970s to 6.5 billion people in 2005. In just 35 years, world population increased by more than 3 billion people – a quantity more than the growth in the previous 35 thousand years! Second, the dramatic economic growth that took place in the last century, leading to the process of globalization of the economy. As remarked by the Millennium Ecosystem Assessment,[2] technological progress has been able to handle pretty well this dramatic increase in the size of human societies:

> *Since 1960, while population doubled and economic activity increased 6-fold, food production increased 2½ times, food price declined, water use doubled, wood harvest for pulp tripled and hydropower doubled.*[3]

And third, an increasing stress on the environment and natural resources, which has been generated by the simultaneous skyrocketing of both population and affluence. Again quoting the findings of the Millennium Ecosystem Assessment:

> *Over the past 50 years, humans have changed ecosystems more rapidly and extensively than in any comparable period of time in human history. This has resulted in a substantial and largely irreversible loss in the diversity of life on Earth.*[4]

In considering the combined effects of these changes, it is clear that the economic problem associated with the need of satisfying the rapidly increasing demand for energy while respecting the environment is more and more becoming a mission impossible. In fact, if we admit that increased consumption of natural resources associated with an increasing consumption of energy, specifically fossil fuels, is required to produce and consume more goods and services per capita for more people, then we have also to admit that, sooner or later, economic growth will have to face the unavoidable existence of biophysical constraints. As Daly has stated, economic growth nowadays is taking place in a 'full world' (Daly, 1996).

In the field of energy policy, the discussion has been dominated by the debate of two key issues: first, peak oil (since fossil energy is not renewable – it is not produced, but extracted from stocks – the finiteness of these stocks and the continuous increase in the pace of consumption entails that sooner or later the reserves of oil and natural gas will be depleted). Peak oil indicates the point on the curve at which the pace of discovery of new reserves becomes lower than the pace at which existing reserves are depleted. This is analogous to a situation in which a bank account starts to be depleted because spending (withdrawal) surpasses earnings (deposit). And second, the global warming associated with the greenhouse effect (the accumulation of CO_2 and other gases generated by the metabolism of the global economy is affecting the normal functioning of Gaia[5]).

There are two ways to deal with the huge predicament associated with the acknowledgement of the unavoidable existence of biophysical constraints affecting the feasibility of 'perpetual economic growth' on a finite planet:

1   considering the option that humans should start looking for alternative patterns of development no longer based on the maximization of GDP; or
2   remaining tied to the ideological statement that the exponential growth of both population and consumption per capita can go on for ever thanks to a continuous supply of 'silver bullets' provided by technological progress.

As a matter of fact, traditional economic theory suggests that the problem represented by the existence of biophysical constraints on the expansion of the global economy will be solved by the markets. The theory states that:

> *as demand for energy increases and the supply of natural resources to produce energy diminishes, the price of energy will increase. These price signals will encourage investment in energy-efficient technological advancements.* (Hicks, 1932, pp124–125)

Policymakers around the world have clung to Hicks's 'induced innovation' hypothesis and made it one of the central components of their national energy and environmental policies. Technology will create environmental improvement with the least effect on the economy (Foster, 2000).

This confidence in the power of progress and technology provides the justification for refusing to consider the hypothesis of looking for alternative paths of development. Indeed the great achievements of human progress (the first quote of the Millennium Ecosystem Assessment) seem to justify such ideological intoxication. In the 20th century, technological progress provided humankind a power never dreamed of in the past. However, this technical progress was driven by a huge increase in energy consumption – fossil energy. In other words, since the industrial revolution the success of human technology has depended on a continuous increase in the rate of consumption of fossil energy. What then would happen if fossil fuels run out?

This is why, to save the dominant civilization from a possible decline or from the stress of re-discussing existing priorities, many believe that another type of silver bullet is needed: 'energy-efficient technological improvements'. This, it is claimed, is the type of solution that progress has to provide against energy shortages and, by extension, environmental degradation. Indeed many policymakers, traditional economists and members of the general public believe in this solution. With this book we want to challenge this belief.

Certainly, one would think that improvements in energy efficiency will reduce energy consumption and increase the effect of a given supply. Yet the point we want to make in this book is that this is not always the case. We aim to show that increased energy efficiency leads to increased demand and consumption of energy. This hypothesis is an extension of the Jevons Paradox, which operates when an increase in efficiency in using a resource leads to a medium- to long-term increase in the consumption of that resource rather than a reduction (Giampietro and Mayumi, 2006).

As noted earlier, the growing appetite for energy is a product of many factors, most notably rising incomes, increasing population, better access to energy and increasing international trade. Thus the implications of the existence of the Jevons Paradox in the energy sector are numerous. Primarily, the Jevons Paradox would indicate that market-based solutions will not solve today's energy or related environmental problems. Currently over half of the world's population do not have access to commercial energy (Banerjee, 2005, p2). As those countries that are on the verge of developed country status continue to modernize, the demand for energy will further increase. Thus the energy/environment nexus is at a critical stage.

Nearly all the products consumed in the world today are produced using fossil fuels, which are not a renewable source of energy. The traditional school of thought would have you believe that technological improvements making energy use more efficient will be the solution. However, as you will see throughout this book, this is not the case. The Jevons Paradox is little known outside some academic circles, but we argue that a sound understanding of it is important not only for policymakers but also for other stakeholders and the general public.

We have written this book to provide a warning that relying on energy efficiency and technology as a solution is foolhardy. The book is organized so as to provide a complete introduction to the Jevons Paradox, from its origins, through a theoretical framework of the topic, to an applied empirical approach.

Chapter 2 provides a detailed historical background of the Jevons Paradox and frames the issue from a historical perspective. In it Blake Alcott provides a thorough and detailed introduction to the topic. Specifically, he reviews the historical roots of the Jevons Paradox, examines the theoretical case for the Jevons Paradox and then applies that to the modern version, known as the rebound effect. Alcott then explores Jevons's analogy with the employment effects of improved efficiency of labour and presents an analysis of these arguments. Finally, he debates how to incorporate Jevons's findings and the rebound debate into sustainable development policies.

Chapter 3 examines the issue from both an epistemological viewpoint and a thermodynamic viewpoint and then offers an alternative to the traditional economic method of modelling the interaction between the economy and the environment. Mario Giampietro and Kozo Mayumi build upon the background provided in Chapter 2 to present the epistemological challenges of modelling evolving metabolic systems. This chapter also presents a thermodynamic analysis of the Jevons Paradox. Such a discussion is important because social systems are evolving open systems that cannot escape the constraints set by the laws of thermodynamics. Within this larger whole, the energy market is just a subsystem. The main purpose of the chapter is to provide a general theoretical framework by which a comprehensive understanding of the paradox can be acquired. At the same time it is argued that the Jevons Paradox reflects a standard epistemological predicament associated with the analysis of evolving metabolic systems organized in nested hierarchical levels – social and ecological systems are typical examples of these systems. The Jevons Paradox is always with us when perceiving, representing and analysing these systems.

Chapter 4 provides an empirical analysis to provide evidence that the Jevons Paradox may exist at both national and regional levels. John M. Polimeni uses an analysis of various countries and regions to provide empirical evidence that the paradox may operate at a macro level for energy consumption. Specifically, he uses some of the primary variables thought to cause increases in energy consumption, as well as a proxy for energy-efficient technological improvements, to decipher whether energy efficiency is the primary factor in increased energy consumption. Previous empirical studies have shown that the Jevons Paradox operates for individual energy consumption uses or types of energy, but few have explored it from a macroeconomic perspective. Analysis of this kind is important because policymakers are relying on technology to counter the effects of increased energy demand, and hence increased consumption of natural resources.

The book ends with a summary of the findings and a discussion of the implications of the Jevons Paradox. Included in Chapter 5 is an examination of alternative energy policies that may be used to counter the path the world is on by relying on energy-efficient technologies as a solution.

NOTES

1 www.census.gov/ipc/www/wp98001.html.
2 Finding number 2 in the executive summary slide show presentation (see www.millenniumassessment.org/documents/document.360.aspx.ppt).
3 Finding number 1 in the executive summary slide show presentation (see www.millenniumassessment.org/documents/document.360.aspx.ppt).
4 www.millenniumassessment.org/documents/document.360.aspx.ppt.
5 The expression 'Gaia' refers to the conceptualization that the planet Earth should be viewed as a complex of autopoietic systems acting as a sort of integrated super-organism. This idea was proposed originally by Lovelock and Margulis (1974) and then elaborated in more detail in Lovelock (1979). The name Gaia refers to the concept of Mother Earth and it was used by the Greeks for indicating the relative goddess.

REFERENCES

Banerjee, B. P. (2005) *Handbook of Energy and the Environment in India*, Oxford University Press, New Delhi

Daly, H. (1996). *Beyond Growth: The Economics of Sustainable Development*, Beacon Press, Boston, MA

Foster, J. B. (2000). 'Capitalism's environmental crisis – Is technology the answer?', *Monthly Review*, vol 52, no 7, pp1–13

Giampietro, M. and Mayumi, K. (2006) 'Efficiency, Jevons's paradox and the evolution of complex adaptive systems', in A. Sinha and S. Mitra (eds) *Economic Development, Climate Change and the Environment*, Routledge, New Delhi, pp203–223

Hicks, J. (1932) *The Theory of Wages*, Macmillan, London

Lovelock, J. E. (1979) *Gaia: A New Look at Life on Earth*, Oxford University Press, Oxford

Lovelock, J. E. and Margulis, L. (1974) 'Atmospheric homeostasis by and for the biosphere: The Gaia hypothesis', *Tellus*, vol 26, no 1, pp2–10

2

Historical Overview of the Jevons Paradox in the Literature

[In] a stationary condition of capital and population ... the industrial arts might be as earnestly and successfully cultivated, with this sole difference, that instead of serving no purpose but the increase of wealth, industrial improvements would produce their legitimate effect, that of abridging labour. Hitherto it is questionable if all the mechanical inventions yet made have lightened the day's toil of any human being. They have enabled a greater population to live the same life of drudgery and imprisonment, and an increased number of manufacturers and others to make fortunes. (Mill, 1848, pp756–757)

INTRODUCTION

For William Stanley Jevons's immediate predecessor John Stuart Mill, according to the above epigraph, the legitimate effect of 'industrial improvements' such as efficiency increases would be less work per capita. This is, after all, *enabled by* labour efficiency increases at the same level of affluence. In the same manner, today's environmental strategy of technological efficiency holds that the legitimate effect of energy efficiency improvements is less energy consumption at the same or an even higher level of affluence. Jevons asked, and to his satisfaction answered, the question of whether energy efficiency by itself leads to this hoped-for result or whether it leads to a higher rate of energy resource consumption. He titled the seventh chapter of his 1865 book *The Coal Question* 'Of the economy of fuel', which confronts us with the 'paradox' that less fuel consumption per unit of equipment causes greater total consumption (p141). Fuel can be 'saved' per unit while not at all being 'spared' for posterity's use (p155).

The fuel in question was the coal to which Britain owed its affluence, power and civilization; the worry was that supplies, especially easily mined ones, were dwindling fast. Some experts advised not to worry because coal's use in steam engines, smelting and so forth was becoming more and more efficient, a view to which Jevons objected by means of his 460-page argument that 'it is the very economy of its use which leads to its extensive consumption' (p141). And while today's fuel worries concern pollution somewhat more than depletion, the paradox remains. Why otherwise would virtually all governmental bodies, green lobby groups and the greater part of public opinion favour efficiency increases to reduce our rate of overall consumption? Yet many academics take Jevons's part in doubting this.

To his brief statement of his thesis Jevons cheekily added:

> *Nor is it difficult to see how this paradox arises. ... It needs but little reflection to see that the whole of our present vast industrial system, and its consequent consumption of coal, has chiefly arisen from successive measures of economy.* (pp141–142)

Today, however, the *solution* of the paradox is requiring a great deal of reflection, of which the present book is a part. The revival of Jevons's argument by Leonard Brookes (1978 and 1979) and Daniel Khazzoom (1980), both of whom doubted the environmental efficacy of the efficiency standards for cars, refrigerators, houses and light bulbs that were being enacted in the decade that saw the Club of Rome report[1] and OPEC fuel price hikes, opened a heated debate. In Khazzoom's words:

> *... changes in appliance efficiency have a price content ... with increased productivity comes a decline in the effective price of commodities, and ... demand does not remain constant ... but tends to increase.* (Khazzoom, 1980, pp22–23)

While this new/old insight that efficiency increases trigger *some* additional input consumption – known by the cute technical term *rebound* – was readily acknowledged by all, a school of thought emerged regarding it as 'insignificant' (Lovins, 1988, pp156–157) or 'small' (Schipper and Grubb, 2000, pp367–368 and 394–386), meaning that greater efficiency would indeed bring net resource savings. Empirical attempts to measure economy-wide rebound have failed, however, and theorists have indecisively argued the pros and cons of Jevons's extreme and very important thesis that rebound is not only significant but in truth greater than the savings theoretically possible when equipment becomes more efficient and demand stays constant.

Rebound of more than 100 per cent of theoretical 'engineering savings' is called *backfire* because in this case environmentally motivated efficiency measures are counterproductive. As we will see, Jevons's economist predecessors made Khazzoom's point of rebound's necessity in countless passages in their treatises on the principles of political economy. Concerning Jevons's backfire thesis, however, they were largely silent: the question had not yet arisen. Nevertheless, some of their time-tested insights can aid today's search for a definitive answer to how much more energy consumption results from greater energy efficiency – an assistance sorely needed in a debate plagued by rudimentary difficulties of definition, taxonomy and methodology (Sorrell and Dimitropoulos, 2006).

Some of the open questions are as follows:

- What would a proof for or against backfire even look like?
- What is the strict definition of rebound? Of what, exactly, is it a percentage?
- What is energy efficiency? While energy inputs are perhaps easily defined and measured, with what outputs are they to be compared? Are these in physical, monetary or welfare units?
- Do we even need the concepts of theoretically possible savings, rebound and backfire, or can we, for example, describe a production function then note that if a factor such as energy becomes relatively more productive, demand for it goes up more than it would have otherwise?
- Can we fully trace consumers' reactions to their increased purchasing power (income effect) resulting from lower prices?
- Can we, for instance, measure efficiency elasticities of price and then price elasticities of demand for both the goods and services and the primary energy inputs themselves?
- Many approximations exist for *direct* rebound, in other words the energy consumption increase entailed by increased consumption of goods and services produced more energy-efficiently. But what about *indirect* rebound for other products that now fall within the budgets of many consumers?
- Is macroeconomic empirical work – regression analysis with energy consumption as the dependent and energy efficiency as an independent variable – even possible? (see Chapter 4)
- At what scale is such work fruitful? Are studies limited to sectors, countries or groups of countries (usually OECD) helpful?
- Can standard models of energy consumption continue treating population size and GDP as wholly exogenous, or are they partly a function of energy efficiency?
- Can we assume that human beings will continue to multiply and consume rather than take 'efficiency dividends' in the form of less reproduction, work and production?
- What is the experience of the last three centuries with increasing *labour-input* efficiency? Have these caused less population and employment, in other words was rebound less than 100 per cent?[2]

Discouraged by this state of affairs in rebound research, I took inspiration from the title of Jevons's first chapter, 'The opinions of previous writers', and turned to the classical political economists. To be sure, the writers Jevons surveyed by name were not the 'old-timers' of political economy, but rather geologists, politicians and mining engineers. Nevertheless, it seems clear that it was the economics texts of the 19th century that gave Jevons so much confidence in his thesis and that discouraged challenges by later economists.[3] By *The Coal Question*'s posthumous third edition of 1906, petroleum had certainly taken the pressure off coal, just as coal had taken the pressure off wood (Jevons, pp183–185; Hearn, 1864, pp194–195), but how could succeeding economists resist the chance to wrestle with a paradox unless the consensus saw the question as settled?[4] For Thorstein Veblen, for instance, it was sure knowledge that latent demand would lap up every efficiency gain (1899, pp32, 110 and 241), and Harold Hotelling wrote that the goal of resource conservation, traditionally, was pursued by either proscribing production or prescribing inefficiency (1931, p137).

With due respect for the efficiency conundrum – how can per-unit efficiency be outweighed by the sheer number of consumed units? – but with the reassurance that a paradox is only an *apparent* contradiction, let us examine the main works of William Petty (1675), Richard Cantillon (1755), Adam Smith (1776), Jean-Baptiste Say (1803), Lord Lauderdale (1804), David Ricardo (1817), Jean Simonde de Sismondi (1819), Thomas Robert Malthus (1820), John McCulloch (1825), Richard Jones (1831), Charles Babbage (1832), John Rae (1834), John Stuart Mill (1848), William Hearn (1864) and Karl Marx (1887).[5] Jevons mentions, and extremely favourably, only Babbage, Mill and Hearn, but all dealt explicitly with efficiency and named it as a cause in their explanations of the increases in population and wealth so palpable in Europe and North America. Efficiencies of varied provenance were increasing: of the individual labourer, of the organization of production, of the institutions of society, and of the technology of using tools, mills, machines, energy and materials, the last constituting Jevons's and our realm of interest. Although for them the increase in demand for labour, land, coal and metals was no less palpable, on our question of whether this increase in wealth entailed an increase in consumption of these *inputs* to wealth, they shed only indirect light. Yet because their and Jevons's analyses contain all the concepts in today's debate, they offer a chance to clear up our thinking. To be sure, today's bone of contention – whether greater consumption of inputs is *due to* (Brookes, 2000, p356; Moezzi, 2000, pp525–526) or *despite* (Howarth, 1997, p3; Schipper and Grubb, 2000, p370) efficiency increases – was not buried for us until Jevons's book of 1865.

Our 'previous writers' did, however, close in on the gist of our subject in their lengthy debate over *labour* as opposed to *energy* efficiency. Alongside energy, space and materials, no production can do without the input of working hours, and it was indeed in terms of labour productivity that 'progress' in the 'arts' of agriculture and manufacture was defined, as when Jevons refers to the labour-saving invention of gunpowder (p105). Their examples of the making of pins, books, stockings, metal and flour were expressed in terms of output per worker or per man-hour, and analogous to energy inputs one could and did argue that such 'progress' meant unemployment. In his curt rejection of this argument (p140), Jevons was standing on an explicit controversy involving not only Luddites, Owenites and industrialists but also Say against Sismondi and, with more ambiguity, Malthus and McCulloch against Ricardo (also later Marx, Part IV, Chapter 15). Note that in terms of today's debate, the position taken by Sismondi that work efficiency causes less total work is analogous to today's position that energy efficiency effects a rebound of less than unity: unemployment, that is, of either labour or fossil fuels. If labour inputs are really saved, *ceteris paribus*, by increasing the efficiency of their use, then any growth in work hours (including population) must be due to other factors. The contrary position, taken by Say, holds that those immediately and distressingly laid off will find work, albeit usually not in their former occupation. This is 'backfire': saving work per unit creates more work overall – our paradox.

This chapter is not organized chronologically but according to concepts and arguments used in today's debate. Statements by the 'old-timers' are enriched with references to similar contemporary positions. The categories are:

- What is output/input efficiency?
- How is the output numerator defined?
- Do efficiency increases cause wealth increases?
- How does efficiency change affect prices and profitability?
- Do efficiency increases amount to a societal free lunch?
- Is rebound proven?
- Do consumers choose further consumption or indolence?
- Is backfire proven?
- How do we deal with population growth?
- Is there technological unemployment?
- What would resource and labour consumption be if technological efficiency had not increased?

Jevons's own conclusions and arguments have been analysed previously (Alcott, 2005) and are here spread throughout the text.

Please keep these methodological points in mind:

- We are asking whether lower energy or labour inputs per unit of 'product' *cause* lower input consumption economy-wide; our independent variable is thus a ratio. Our dependent variable, on the other hand, is a total or absolute amount, namely of resource depletion or emissions – the values of interest to the environmental problem since, metaphorically speaking, the environment does not 'care about' *ratios* of outputs and inputs or of consumption or pollution per person or per unit of GDP or per rich or poor nation.[6] The formal problem confronting all rebound measurement is that it is impossible to derive an absolute number from a ratio or change in a ratio; without further factual information, an 'extensive' number cannot be deduced from an 'intensive' one (Giampietro and Mayumi, 2000, pp183–187 and 191 and Chapter 3 of this volume).

- Must we seek *necessary* connections? In our case this would involve assumptions regarding human nature and the particularities of human societies, mainly whether or not consumers, including marginal ones, are saturated. Absolute saturation regarding all goods and services would mean a rebound of zero; the income effect would disappear because people would choose to earn and spend less and theoretical 'engineering' savings would equal real savings. But with any positive price elasticity of demand we have some additional consumption. Thus we must always compute or judge the probability that consumers will keep doing more-or-less what their parents did (Jevons, pp192–196).[7]

- A worldwide regression analysis would have to include data on energy efficiency, energy consumption and energy prices. The last two can be traced with some certainty,[8] but, as we shall see, efficiency presents severe data and definitional difficulties. Since products and activities come and go, over time the 'output' part of our ratio is a moving target (Rosenberg, 1982 and 1994; see also Chapter 3). Must we resort to that lame workhorse GDP, or can we find *physical* output metrics like 'useful work' or 'exergy' or tons or volumes, perhaps unaggregated? We would also have to control for other factors like non-technological efficiency increases[9] and partially exogenous population and wealth.[10] Nevertheless, few would deny that technological efficiency has increased, and regression analysis offers undisputed insights (see Chapter 4).

- *Direct* rebound is a pet subject of study, but in and of itself is not relevant for environmental policy, which needs to know economy-wide rebound adjusted for trade of embodied energy. If nevertheless computed, researchers owe us a demonstration of *how* to use it in calculating *total* rebound. At the minimum, the ambiguity in much of the literature as to which rebound is being discussed must be eliminated (Greening et al, 2000, pp390–392; Berkhout et al, 2000, pp425–431).

Please recall the urgency of this policy question. Depletion concerns seem today perhaps unimportant, although they remain both inexorable and ethically binding. Among Jevons's many emotional passages are those where he attests the 'religious importance' of the coal question, where he laments living off 'a capital which yields no annual interest' or where he quotes Drayton concerning the fuel voracity of the iron industry: 'These iron times breed none that mind posterity' (pp14, 412, 373 and 136). Moreover, Jevons advocated using coal-given prosperity for posterity and for a sort of soft landing at coal's limits (ppxlvi–xlvii, 4, 37, 156, 184, 195, 200, 232, 274–275 and 455; Boulding, 1966). Running out of fossil fuels can, however, be spread over a long time horizon or ameliorated by using them as embodied energy in renewable energy installations. But two other sets of concerns stand no postponement: first, and obviously, our present and intensifying planetary greenhouse with its welfare consequences; and second, and today óften ignored, the side-effects of the machines and infrastructure that enable and embody energy efficiency: noise, accidents, public ugliness, local air pollution, overuse of fresh water, monotonous work, and so on. The community of ecological and environmental economists should waste no more time in delivering a decisive, policy-useful *judgement* on this question: is efficiency part of the solution or part of the problem?

WHAT IS EFFICIENCY?

Like all cost-cutting efficiency increases, energy efficiency until recently exclusively served the goals of higher profits and greater average affluence. In so far as the costs of the efficiency introduction itself could be amortized, they are the business-as-usual maximization of material well-being. This fact is today often downplayed or ignored when energy efficiency increases are singled out to serve the contrasting *environmental* goal of lowering the yearly rate of energy consumption and/or pollution. In whichever way they are perceived, though, they are the starting point and logical centre of our investigation. As such they warrant careful definition and taxonomy.

Throughout the following examination of our authors' definitions of efficiency it is axiomatic that efficiency denotes a *ratio*. The numerator is output and the denominator is (energy) input. 'Efficacy', 'effectiveness' or, more ambiguously, 'power' denote in contrast the causation of a given amount of output regardless of cost or input. Ontologically, the thing that is more or less efficient is the input. In classical parlance, *power* resided in the inputs of labour and nature, measurable in terms of what a certain amount of these could *produce*; the classical production function was $Q = f(\beta M, M, \alpha L, L)$, where M was material/energy, L was labour and the Greek letters were productivity coefficients.[11] The ubiquitous classical concept of 'productive power' thus implies, like the Latin-based term efficiency, both a 'making' and an 'out of something'. The inverse of efficiency is *intensity*, as in the 'material intensity of production' common in today's environmental efficiency discussion (Schmidt-Bleeck, 1994; Hinterberger et al, 1997; von Weizsäcker et al, 1997). The ratio describes, moreover, the amount of input *per unit* of output. Finally, we are not investigating consumption efficiency – for example boiling only the amount of water needed for the cup of coffee (Hannon, 1975, p96; Etzioni, 1998, p630; Prettenthaler and Steininger, 1999; Norgard, 2006).[12]

Of a certain area of land William Petty asked, 'How many Men will it feed?', implying an output/input ratio of food over square metres and holding food per Man constant; he offered data on the agricultural productivity of 'improved Acres' (1676, pp286–288). Cantillon likewise employed this agricultural paradigm either as rice/m^2 or as yield/seed (1755, pp26 and 128). Departing from the spatial metric, Petty also attested differences in transport efficiency for 'bulkey Commodities' between 'Water Carriage' and 'Land Carriage', a given output of bulk times distance achieved by less (water) or more (land) input of time as well as endo- or exosomatic energy (pp255 and 293–294). Using the examples of flour grinding and printing, his 'Arithmetick' showed, for instance, that a mill after deducting the labour embodied in its construction 'will do as much Labour, as Four Men for Five Years together' – an efficiency increase of 20 times; with printing a factor of 100 results; the wagon means that 'one Horse can carry upon Wheels, as much as Five upon their backs' (pp249 and 256).

Petty's endeavour is to explain why different European nations of similar size and population have different levels of wealth. Like Malthus (1824, p265), Mill (p100) and Solow (1957), his *explicans* turns out to be not such absolute quantities of land or people but their productivity ratios: England was more efficient and therefore richer than France or Holland. Would that we could today use the method of Petty and Solow for our *explicandum* of energy inputs.[13] But unless we can take GDP as a good proxy for output, this path is closed to us: both the 'dematerialization' of GDP and the difficulty of identifying what it is that GDP measures constitute major difficulties. A godsend would be a time-series of two non-trading countries similar in all respects except level of technological efficiency.

Presaging today's computations of theoretical 'engineering' savings, Petty even reckons the monetary savings from innovations (pp255–257). In other words, costs of production fall and society, left with at least the same amount of flour, printed matter or transport as before, would have purchasing power left over.[14] Petty explicitly attests huge labour savings (pp306–308), but his only remark bearing on labour rebound is that as a result of 'improvement' of 'Art' many millions *could* work, but aren't 'disposed or necessitated to labour' (pp249 and 307). This hints at a normative issue that confused the discussion between Say, Sismondi, McCulloch, Mill and Marx: given that work is basically a painful, irksome cost, 'unemployment' would be a good thing, and, like today regarding energy inputs, we should hope for low or no rebound.[15] But in the absence of political means to spread work equitably, by bestowing purchasing power work becomes a good thing.

As his title and introduction reveal, Smith's *explicandum* was wealth or 'produce', usually defined materially (I.v, I.viii.21, IV.ix.38 and V.ii.e.10).[16] His favourite explanatory variable was the intensive one of 'productive Powers [of Labour]', itself mainly explained by a number of variables, including division of labour, dexterity, work organization and machines, themselves explained by the 'propensity in human nature ... to truck, barter and exchange one thing for another' (I.i and I.ii.1). The only other factor raising production is an increase in labour's *quantity* (I.intro.3–4, II.iii.32 and IV.ix.34–36). Productive *power* is 'the quantity of work [*produce* such as nails], which ... the same number of people are capable of performing' and its increase is 'improvement' (I.i.5 and I.i.6). Surrounded by increasing population and production, it is not surprising that Smith does not define efficiency the other way around as a constant output with less input: the fact was that *number of pins* rises (by a factor somewhere between 240 and 4800), not that society spends fewer hours making pins (I.i.3). Smith also framed productivity in other terms, attesting, for example, the greater efficiency of water over land transport, his ratio being that of tons 'carried'/man, and, as with his pins, the waters between London and Leith are plied more often (I.iii.3 and I.xi.b.5). Jevons later showed that canals lowered coal prices, a case of greater *transport* efficiency raising *coal* consumption (pp121–122 and 166).

Smith's denominator was sometimes space (land, soil), with output as food or wool (I.xi.b.2–6 and 15, and 1V.ix.5–6; see also Say, p295), and sometimes mines (thinkable in m^3) of varying 'fertility' (I.v.7 and I.xi.c.10–11). The productivity of the soils and mines in turn partially determine the efficiency of labour. Again, quantity is a function of both the productivity and quantity of the material itself and of labour (with capital able to increase both productivities). In Say this material factor is the *agens naturels* or *services productifs*, with 'agency' denoting the 'power' and the power's strength determining the agent's 'fertility' or 'fecundity' – here with no reference to labour inputs (pp40, 63–77, 101, 127, 301 and 395). Jevons similarly asserted that 'power' was 'in' coal – and that it was power that had through 'increased ... efficiency become *cheap*' (pp145–146 and 186). In contrast to later neoclassical neglect of material as a productive factor, he held that 'in our successes hitherto it is to nature we owe at least as much as to our own energies' (p318). Similarly, coal and oil, as well as coal mines and oil 'fields', have varying inherent fertility in both chemical terms and terms of ease of access. Ricardo confirmed this ambiguity in the concept of material efficiency by noting that 'improvements in agriculture are of two kinds: those which increase the productive powers of the land, and those which enable us, by improving our machinery, to obtain its produce with less labour' (p80; see also Smith, I.xi.d.1; Mill, pp724–725).

As the pin and nail examples show, Smith by no means neglected manufacturing, for example the 'woollen manufacture', where the 'working up' of a 'quantity of materials' was facilitated by 'a variety of new machines' (I.xi.o.12 and II.intro.3). But his usual denominator was labour input (I.ix.34–35 and I.xi.4): for land of *given* fertility, then greater produce results only from the greater 'efficacy of human industry [= *labour*, not manufacture], in increasing the quantity of wool or raw hides' (I.xi.m.14). Note especially that often 'improvement' was expressed as *less* labour input for 'any particular piece of work' (I.xi.o.1). This holds output constant and is the version of the ratio found in Ricardo, for whom 'economy in the use of labour' or labour's 'abridgement' – by means, for instance, of engines – meant lower or at least not higher 'charges of production' (pp25, 26, 41, 69 and 397). But more often Smith's ratio change held input constant against a 'great increase of the quantity of work [= *produce*, not labour]' (I.i.5); with good farm capital and the 'best machinery', the same amount and quality of labour made a 'much greater quantity of work' (II.ii.7; I.viii.3 and I.xi.o.12).[17] Malthus's rendering of efficiency change likewise described 'a machine in manufactures ... which will produce more finished work with less expenditure than before' (p145).

As with the question of whether a glass is half full or half empty, it matters whether we define efficiency increase as 'less input per unit of output' or 'more output per unit of input'. Although technically equivalent, the former biases our thinking by holding output constant and looking at what could be saved while the latter biases it by highlighting increased output with perhaps no saving. A simple example is replacing an open fireplace with a ceramic stove: one can heat the same amount of space to the same temperature, thus really saving firewood, or use the same amount to heat more rooms warmer.[18] Starting one's chain of thought with resources still available for more economic activity (after they are rendered able to produce more per unit) is conducive to perceiving large rebound; in Hearn's words, greater efficiency 'sets free a quantity of commodities ... or ... materials' (p271).

Say's denominators were both labour and materials like land, water, mines, wind and other *agens naturels*. In some cases 'tools and machines ... enlarged the limited powers of our hands and fingers'; in China tools for 'drilling, in lieu of the broadcast method [of sowing], raises the productivity of land' (pp86 and 394). In other cases 'useful machinery' is 'strengthening and aiding the productive powers of nature', the category within which today's energy efficiency efforts fall (p357). He insisted on the equivalence of ratios with higher numerators (output) and those with lower denominators (input):

> Every saving in the cost [les frais] of production implies the procurement, either of an equal product by the exertion of a smaller amount of productive agency [$Q_{same}/expense_{less}$], or of a larger product by the exertion of equal agency [$Q_{more}/agency_{same}$], which are both the same thing. (Say p301; see also pp86, 88, 201, 204 and 395)

However, while he sometimes thus underlines the 'saving of productive agency' (p395), Say's excitement is aroused by the opposite case, namely 'to obtain a larger produce from the same quantity of human labour. And this is the grand object and acme [*le comble*] of industry' (p86).

Note that one of his examples describes an increase of labour efficiency (αL) whereby one man mills as much as ten men previously when a windmill by means of sails (capital or K) is substituted for a treadmill (pp74–75).[19] While this is clearly an increase in labour efficiency, a case of 'capital enlarging productiveness' (p77), it is *not* an increase in *wind* efficiency (βM) – unless starting from zero. Similarly, the first internal combustion engine did not increase the economy of fuel but only the economy of transport in terms of time and labour. Therefore, innovation seems not always subsumable under efficiency. Say does hint at a distinction between an invention – effecting the first-time use of a natural resource – and a new 'process' to 'produce ... an old [product] with greater economy', for example a new 'method of reducing the friction of bodies' (pp329 and 433).[20] Another, endearing example was the use of sulphuric acid to destroy the 'mucilaginous articles of vegetable oils', which could then be substituted for expensive fish oil, an efficiency increase, in the broadest sense, that 'placed the use of those lamps ... within the reach of almost every class' (p116). Here the production of lumens became more efficient, but not that of vegetable oils in producing lumens, because these were not before used. Brindley, on the other hand, observed that the Newcomen engine wasn't efficient enough for coal to replace 'the power of horses, wind or air' (Jevons, p143). This seems to be a case of increased efficiency in the use of an exosomatic energy source, already stutteringly in use, substituting for others whose efficiency potential had been exhausted.

In discussing rebound we should take this distinction between innovation and technological efficiency seriously: when cutting tools change from steel to ceramics to carbide (diamonds) these raise cutting efficiency but are not more efficient uses of a given material (Rosenberg, 1982, pp3–4 and 65). Malthus's more abstract formulation distinguishes between the invention of machines and the more efficient or 'best' machines replacing less efficient ones (pp145, 170 and 229). With Rae the distinction is straightforward – between 'new arts' and 'improvement in the arts already practised' (pp15; see also 224 and 253). His examples include the plough itself as opposed to better ploughs, macadamized as opposed to stone roads, and better steel tools (pp87, 114, 226–228 and 259). He moreover traces the steam engine's invention, improvement and connection with coal mining in terms almost the same as Jevons's (Rae, pp245–248; Jevons, pp142–153; McCulloch, pp97–99).[21] Hearn wrote that:

> By ['improvement'] I mean not the discovery of natural agents previously unknown or unused; but the knowledge of new combinations of agents already known. ... Those improvements which increase the efficiency of the actual agent [coal] are ... distinct from those inventions the utility of which consists in the abridgment of human labour, and the substitutions for it of physical forces. (pp99–100)[22]

First, for instance, India rubber was used to do new things, then it became more efficient through vulcanization and sulphur treatment; coal likewise was first found and substituted for charcoal, then made more efficient through the hot blast in smelting (Hearn, pp100–102).

The point is that greater resource consumption caused in the first place by inventions should not be booked under rebound. That said, Malthus has a point that inventions 'are the natural *consequence* of improvement and civilization' (p281). In other words, efficiency increase can cause inventions and new uses.[23] At any rate, once more, identifying which efficiency changes to measure is vexed both by new products and by better-'quality' products that may even constitute efficiency *decreases*. Rae lamented that while of course 'wealth' had vastly increased since Henry VII, there had been 'not only an increase, but a change' (pp18–19; see also Chapter 3).

For 'efficiency' Malthus uses not only 'productiveness' and 'fertility' but also the 'facility' or 'difficulty' of producing or obtaining output, again almost always in terms of labour input. At times he emphasizes 'saving of labour' or 'relief from labour' in producing 'a given effect' (pp128, 152 and 170), at times a greater production (pp281–283; Malthus, 1824, p63), and once simultaneously greater 'finished work' with 'less expenditure' (p145). Referring to Say, who had written that 'a landed estate may be considered as a vast machine for the production of grain, which is refitted and kept in repair by cultivation; or a flock of sheep as a machine for the raising of mutton or wool' (Say, p86 note, p318 note), Malthus writes:

> *The Earth has been sometimes compared to a vast machine, presented by nature to man for the production of food and raw materials; but, to make the resemblance more just, as far as they admit of comparison, we should consider the soil as a present to man of a great number of machines, all susceptible of continued improvement by the application of capital to them, but yet of very different original qualities and powers.* (pp144–145; see also pp66, 111, 115 and 168; McCulloch, p278)

Say also repeatedly talked of the 'spontaneous gifts of nature', like air, water, light, fire, gravity, pressure and steel (pp63, 71, 75, 86, 286 and 362), all susceptible to improvements through 'industry' which must 'awaken, assist or complete the operations of nature' (pp63–64, 74 and 86; Smith, II.iii.3).

Undoubtedly impressed both by Say and what he observed in rural Canada, Rae likewise repeatedly described the material factors of production and their 'productive powers' (pp10–12); he saw 'fire and water transformed into our obedient drudges' (p14); our 'instruments ... draw forth stores' of materials and 'improvement in their construction ... put additional stores within reach of the nation' (pp19 and 68); a 'North American Indian' improves a 'wild plumb tree' or dams 'a very scanty brook' (p83). The doctrine perceives an efficiency ratio in that:

> *the knowledge of the civilized man, compared with that of the savage or barbarian, gives him the power of constructing a much greater number of instruments out of the same materials.* (p99)

Just as Petty and Smith had distinguished between the quantity of labour and its productivity,[24] Rae's analysis of 'the action of matter upon matter' separated the 'amount of materials' from 'the efficiency of these materials' (pp112–113) but he is additionally discussing the effect of our 'instruments' on matter's efficiency rather than their greater or lesser *inherent* natural power (pp87–110). 'Instruments' roughly mean capital, in other words anything man-made for the purpose of future production, including fields and even food (in classical terms 'circulating capital').[25]

More than our other authors, Rae thus analyses material rather than labour inputs (p99). He also conceptualizes the *costs* of efficiency, once even defining efficiency as the total production of an instrument (until its 'exhaustion') divided by the cost of making it measured in units of labour; this is 'the ratio of the capacity ... to cost' (pp259; see also 173 and 354–355).[26] Smith had already made the pertinent point that the:

> *expense which is properly laid out upon a fixed capital of any kind is always repaid with great profit, and increases the annual produce by a much greater value than that of the support [depreciation] which such improvements require.* (II.ii.7)

With an example of more durable pots and pans taken from Smith, Rae shows that in spite of (because of?) their 'becoming more expensive articles', they 'augment ... national capital ... with advantage to society' and are 'preferred by good economists' (p21). The relevance of the (energy) costs of energy efficiency to rebound is disputed. One solution is simply to deduct these from the savings theoretically possible during the *operation* of the more efficient instrument – thus lowering the quantity of which rebound is a percentage (Jevons, p446).[27]

Rae also distinguishes between 'efficiently' and 'effectually' (in the sense of merely getting a job well done), as when the threshing machine not only saves labour but separates grain *better than* the flail method (p20). This again raises the question of the changing quality of the output in our numerator. Otherwise Rae's treatment closely follows Say's, for example in emphasizing the equivalence of ratios with lower inputs and those with higher outputs (pp66, 92, 131 and 259) If anything, his bias is towards the latter; adding to manufacturing capital will:

> *effect an increase in the productive powers of the community; that is, they give those powers the capability of producing the same quantity of an article at less expense, which certainly must be allowed to be an increase of them.* (p70)

This language comes close to a description of an outward shift of a community's *production possibilities frontier*. This is the key assertion of and proof of rebound, if not backfire: we are *enabled* to produce and consume more without more effort, time or material. Whether backfire obtains depends then on consumer behaviour or, in fancier language, the efficiency elasticity of demand.

Rae and Malthus, whose *Principles'* last edition appeared two years after Rae's treatise, were describing the phenomenon that is the starting point of our investigation: the human ability to get more out of the same amount of nature. Rae's fellow Scotsman McCulloch had a few years earlier written, in the usual terms, that division of labour 'saves labour', but also that 'the invention and improvement of tools and engines' caused a rise in our variable – 'the quantity of raw materials which the same number of people can work up' (McCulloch, pp96 and 99). His term for output is here materially expressed, moreover in terms of raw material rather than material objects. McCulloch also introduced the method of assuming an overnight economy-wide increase of efficiency and then deriving the consequences (pp166–167; Mill, pp723–725). But whereas today researchers at Strathclyde, Scotland, similarly assume an 'efficiency shock' of 5 per cent (Allan et al, 2006, pp5 and 36), McCulloch's was by a *factor* of ten![28] Say later got rhetorical mileage out of assuming 'that machinery should be brought to supersede human labour altogether' – a labour-efficiency 'shock' of 100 per cent (p88)!

Finally, Mill's characterization of efficiency reminds one of *economic* or 'Pareto' efficiency. His causal chain is from an 'extension of the market' (here exogenous) to more 'division of labour' to 'a more effective distribution of the productive forces of society' (pp87–88 and 281). In a passage quoted by Hearn (p68) the doctrine presented to Jevons was that 'any improved application of the objects or powers of nature to industrial uses enables the same quantity and intensity of labour to raise a greater produce' (Jevons, p106).[29] However, greater consumption is merely enabled: equally enabled is a real saving of labour and material inputs. We choose between them.

Mill's numerous descriptions of productiveness epitomize the classical analysis (pp93, 99, 106, 118, 129, 153–154, 710 and 724).[30] Yet notwithstanding his famous defence of the stationary state (pp752–757), one discerns his preference for the growing economy in his remark that the 'increased effectiveness [efficiency] of labour … always implies a greater produce from the same labour, and not *merely* the same produce from less labour' (p133, emphasis added). He also claimed that 'no one would make or use ploughs for any other reason than … the increased returns, thereby obtained from the ground' (which could pay the plough-maker) (p31). That society as a whole – macroeconomically – could choose the version 'same output less input' is impossible. This reflects the normative position persisting to the present day of the unassailability of economic growth, epitomized by Smith's sentiment that Jevons chose for his frontispiece:

> *The progressive state is in reality the cheerful and the hearty state to all the different orders of society. The stationary is dull; the declining melancholy.* (I.viii.43)

As shown later, Malthus stood alone in objecting that we could indeed choose 'indolence' (pp258, 267–268, 283, 284, 320 and 337). More neutrally, Mill presents his parsimonious theory of production:

> We may say, then, without a greater stretch of language than under the necessary explanation is permissible, that the requisites of production are Labour, Capital and Land. The increase of production, therefore, depends on the properties of these elements. It is a result of the increase either of the elements themselves, or of their productiveness. The law of the increase of production must be a consequence of the laws of these elements. (p154)

These laws enable both extremes: less work and less resource consumption to the full extent of the intensive (per unit) 'engineering savings' (Alcott, 2005, p10); or an increase of production and consumption so great that in the end even more work and material resources are *put into* the economic process. Other laws, of human nature and of desires, consumption and reproduction rather than production, determine exactly where, between these extremes, we end up (Jevons, pp25 and 191–201; Princen, 1999; Sanne, 2002; Alcott, 2004).

WHAT IS OUTPUT?

Energy economics literature offers many terms for our numerator: GDP, units of 'service', goods and services, various physical aggregates, 'product', and, vaguest of all, 'economic activity'. In measuring 'eco-efficiency', Reijnders names five metrics for efficiency: 'a product (such as the automobile), a service (for example transport over a certain distance at a specified speed), an area of need (for example clothing), a sector of the economy (for example energy supply and demand), or the economy as a whole' (1998, p14). Let us distinguish three broad categories – money (GDP), utility and matter:

1   GDP's well-known weaknesses include both ignoring large parts of the economy and valuing some losses as gains (Daly and Cobb, 1989, pp401–455). Specific problems in energy models are elaborated by Rosenberg (1982, pp23 and 55), Jänicke et al (1989, pp14 and 391), Schipper and Meyers (1992, p54), Kaufmann (1992, p54) and Cleveland and Ruth (1998, p35); Smil 'deconstructs' the concept of energy intensity in monetary terms (2003, pp66, 71–78 and 81).[31] This contemporary monetary metric of choice was not available to Jevons and his predecessors.

2   The utility or services concept dominating the rebound literature posits an 'energy service' such as a 'passenger-kilometre'. However, as soon as two people ride in a car, efficiency would then have doubled with no

technological change whatsoever, and when a heavy car replaces a lighter car efficiency would stay the same in spite of a technological change especially relevant to environmental impact. Utility moreover ignores waste, an anthropocentric concept referring to tons of gases and materials; at best, integrating them is a complicated exercise in computing and deducting 'externalities'. Should these be excluded from our numerator, or not? For an incisive account of this concept's difficulties see Ayres (1978, pp50–67). Furthermore, the common concept of 'energy services' is invalid: since *every* service (and good) involves embodied and/or operational energy input, any distinction against 'non-energy services' must be arbitrary.[32]

3   A physical metric (including waste) could be in tons, volume, chemical elements, heat, exergy, work defined in terms of force and direction, or non-aggregated lists of products. Jevons used the metric 'useful work' per pound of coal, expressed in 'foot-pounds', and defines thermodynamic efficiency (pp137–138, 148 and 186).[33] A manageable literature has taken up this challenge, usually with the hope of aggregation[34] and sometimes attempting to integrate physical and utility/monetary metrics.[35] Also, probably all of the technological efficiency changes striven for in efficiency policies are susceptible to physical definition: instead of a 'passenger-kilometre' a ton-kilometre, instead of 'heating comfort' a certain temperature rise in a given volume of space over a given time and instead of a kilowatt hour the amount of primary energy involved. A remaining problem is that due to the first law of thermodynamics output always equals input, leaving us without a ratio! Perhaps only a list of consumer and capital goods (and their utilization rates) remains, and an *aggregated* physical metric is impossible.

After ironically speaking of 'the mass of solid goods and useful services', Joan Robinson sought a non-monetary metric for technical progress, choosing the capital/labour ratio with capital physically measured as the 'value of a stock of goods in terms of commodities' or 'equipment, work-in-progress [and] materials and labour measured in terms of time' (1956, pp19, 65 and 122). She concluded, however, that 'index-number ambiguities' are insoluble (pp64–65 and 115) and that 'economics is the scientific study of wealth, and yet we cannot measure wealth' (p24).[36] The classical economists similarly suffered in defining wealth. Its *genus* was material objects or 'produce' for Smith (I.viii.3–9, 21 and 23, IV.ix.38 and V.ii.e.10), Malthus (pp20–28 and 294) and Mill (pp48–49 and 55). Ricardo also regarded 'riches' in terms of the ubiquitous physical concept of 'necessities, conveniences and enjoyments' (and sometimes 'luxuries' or 'amusements') which had nothing to do with exchange values in terms either of money or other objects (pp275–276). Rae criticized Smith's various definitions and tended to treat wealth and capital synonymously and as physical commodities and instruments (pp387–388, 14, 18, 21 and 171). But all acknowledged some *differentia* specifying their (use or exchange) *value* to us. In Lauderdale's typical phrase, wealth was 'the abundance of the objects of man's desire [including] lands, houses shipping, gold and silver coin, wares, merchandise, plate, furniture, etc' (1804, pp146 and 42; Malthus, 1824, pp29, 258–259). In avoiding Lauderdale's criticism (p152) of Smith's emphasis on durable objects, Mill chose with questionable ontology 'permanent utilities ... embodied in human beings, or in any other animate or inanimate objects' (p48).

If the definition of efficiency must include some quality or value element, let us ponder Say's reaction to his insight[37] that was to become the first law of thermodynamics. He said that we confront a:

> *mass of matter [not] capable of increase or diminution. All that man can do is to re-produce existing materials under another form, which may give them a utility they did not before possess, or merely enlarge one they may have before presented. So that, in fact, there is a creation not of matter, but of utility; and this I call* production of wealth. *[Production is] creation, not of substance, but of utility, so by consumption is meant the destruction of utility, and not of substance, or matter.* (pp62 and 387, emphasis added)[38]

Moreover 'creating matter ... is more than nature itself can do' (p65). More than the others, Say thus emphasized utility rather than goods themselves and posited such a thing as 'immaterial product' (pp62 and 119–124). But he also held that 'the ratio of the national revenue, in the aggregate, is determined by the amount of the product, and not by its value' and never denied that some material was necessary for utility to adhere to: the services of musicians and lawyers, for instance, required their food and education as well as wear and tear on their capital (pp122, 124 and 295; Malthus, 1824, pp258–259; Costanza, 1980)

If we include usefulness in our definition, how do we deal with unwanted objects and waste, both of which affect the environment? While Mill's idea of waste was physical, including 'diving-bells sunk in the sea' and the use of too many horses and men to plough a field (pp8 and 51–52) and Hearn gave the example of close parallel mine-shafts (p208), Rae's chapter 'Of waste' deals with the *economic* inefficiencies of fraud, trade restrictions, transaction costs and so forth – making the point in a very different way that less efficiency means less production and consumption (pp313–319). Among the classical economists there was, moreover, some debate as to whether only anthropogenic objects counted as wealth, or also 'air, water and light' (Say, p63; Mill, pp8 and 153), opening up the water/diamonds discussion over use value as opposed to exchange value and scarcity. Jevons, incidentally, counted waste-reduction as an increase of 'economy' (pp30 and 271–272).

A large contemporary literature thus discusses various metrics for 'environmental' (or energy) efficiency in terms of *desirable* output.[39] The attempt is to abandon purely quantitative measures and introduce the 'quality' of energy, as when exergy is taken to measure input (Ayres and Warr, 2005). Similarly, following a general exposition of energy and its transformations, Jevons offered this definition of efficiency:

> *Now it will be easily seen that the resources of nature are almost unbounded, but that economy consists in discovering and picking out those almost infinitesimal portions which best serve our purpose.* (p163; see also p170)

He elsewhere uses the ratio of 'useful work' to 'power' (pp186–187), thus risking conflation of physical and utility criteria just as Ayres and van den Bergh do when insisting on counting high-entropy 'process waste', the difference between 'work done by the economic system [and] the exergy of all inputs' (2005, p103). For if exergy is already defined anthropocentrically as useful or available energy and can, unlike energy, be destroyed (Ayres, 1978, p52), it itself becomes a (desirable) output. Even taking mass instead of energy in both numerator and denominator, where the output is mass 'embodied in the physical output (finished products)' (Ayres and van den Bergh, 2005, p103) does not escape the fact that to identify 'finished products' we need some anthropocentric criterion.[40]

McCulloch, after acknowledging the law of the conservation of matter, laid down the principle:

> *And hence we are not to measure consumption by the magnitude, the weight, or the number of the products consumed, but exclusively by their value. Large consumption is the destruction of large value, however small the bulk in which that value may happen to be compressed.* (p390)

But can environmental studies ignore what is produced but has *no value*?[41] All oxidized molecules, unless they are recycled by means of further energy inputs, as with carbon sequestration, must count as 'final' output. Space heating can be defined by the time needed for the space to return to (lower) ambient temperature from that desired, but the higher-entropy energy is nevertheless part of output. Lumens rather than 'lighting services' can be measured, but light pollution and heat as a 'by-product' are also output. Steel cannot be made without 'scrap'. While a 'first-law' ratio must be one-to-one, 'efficiency' must be variable, perhaps leaving no way around some concept of utility: we must measure inputs only against the output we *like*. While GDP thus aggregates unsatisfactorily, physical or combined physical/utility metrics have not yet been found.

CORRELATION OF EFFICIENCY AND OUTPUT INCREASE

Whatever 'output' turns out to be, Jevons's immediate predecessor Mill captured the classical conclusion that, formally, productiveness is equivalently lower land/labour inputs and 'increased produce'; that what everyday observation showed was a 'greater absolute produce' or a 'long succession of contrivances for economizing labour and increasing its produce' (pp180, 189 and 706; Smith, I.xi.g.20 and II.iii.33).[42] By 1865 Jevons could write:

> *When we turn from agriculture to our mechanical and newer arts, the contrast is indeed strong, both as regards the numbers employed and the amounts of their products. But the subject is a trite one; every newspaper, book, and parliamentary return is full of it: factories and works, crowded docks and laden wagons are the material proofs of our progress.* (p244; see also pp187–188)

But as Rae lamented, 'all we see is the sum produced by [change], the fact of the increase being more easily ascertained than the manner of it' (p19). Thus, while in dozens of passages all writers previous to Jevons tied increased efficiency to increased product, they seldom formally declared necessary connection. Mill for instance claimed:

> *It will be seen that the quantity of capital which will, or even which can, be accumulated in any country, and the amount of gross produce which will, or even which can, be raised, bear a proportion to the state of the arts of production there existing; and that every improvement, even if for the time it diminish the circulating capital and the gross produce, ultimately makes room for a larger amount of both, than could possibly have existed otherwise.* (p98)

'Room is made', production *possibilities* increase, but there is no claim of universal causality.

Jevons praised Hearn's *Plutology* as 'both in soundness and originality the most advanced treatise on political economy which has appeared' (Jevons, p168 note). Hearn, himself explicitly building on Rae (see, for example, Rae, p260) and Justus von Liebig (1851), described the shift in the production possibilities frontier as follows:

> *It is self-evident, as Mr Mill has observed, that the productiveness of the labour of a people is limited by their knowledge of the arts of life; and that any progress in those arts, any improved application of the objects or powers of nature to industrial uses,* enables *the same quantity and intensity of labour to raise a greater produce.* (Hearn, p68, emphasis added; see also p184)

Jevons then contributed two new thoughts: for 'labour' he substituted 'coal', and he asked the further question of the effects of efficiency not on produce but on input consumption. The doctrine is on the one hand curiously conditional but on the other insistent that growth is impossible without improvement in the 'arts' – a conclusion reached by later growth theorists by statistical means (see, for example, Solow, 1957 and 1970)

Remember that the classical concept of efficiency included individual, organizational and institutional as well as material or technological types, often attested in one and the same passage.[43] Seminal statements of 'economic' efficiency also appear explicitly, wherein what the society does produce is compared to what it could produce given certain natural fertility and technology (Smith, I.ix.15; Say, pp166 and 380; Malthus, pp266 and 304). And although not to my knowledge discussed in classical economics, remember that land and labour inputs are mutually dependent; that is, all terms on the right side of $Q = f(\beta M, M, \alpha L, L)$ influence each other, rendering reduced-form expressions inadequate.

Petty already gave a version of classical 'growth theory' in seeing 'greater consumptions' not only of food but of 'coaches, equipage and household furniture' due to 'improved acres' and population density – and even a growth of postage due to transport efficiency (pp287–305; Smith, I.xi.c.7). Cantillon presaged Malthus's principle of population and the concept of carrying capacity using as examples both people and mice: population followed sustenance, itself a function of land and mine fertility as well as the energy and labour of the population (Cantillon, pp43–44, 46, 62 and 128). Whether his concept of labour was only its quantity or also its efficiency is unclear, but in any case greater population and greater consumption entailed each other. As shown later, this idea that people are also produced – fully conceptualized by later writers – is crucial for the discussion of the Jevons Paradox; models of (energy) consumption or of wealth in general that treat population entirely exogenously necessarily significantly underestimate rebound.[44]

If wealth was 'necessaries, conveniences and amusements' or the goods affording these (Smith, I.intro.1–4, I.v.1 and 9; IV.i.17-18), no writer except Ricardo failed to both attest and laud their *growth*.[45] Rae, for instance, made the empirical claim that the wealth of Great Britain was ten times what is was under Henry VIII (pp14 and 18). Smith saw the gradual spread of 'universal opulence' (I.i.10) or at least 'almost universal prosperity' (I.xi.g.20) and by mid-century for Mill economic growth was axiomatic:

> *Production is not a fixed, but an increasing thing. When not kept back by bad institutions, or a low state of the arts of life [technology], the produce of industry [labour] has usually tended to increase; stimulated not only by the desire of the producers to augment their means of consumption, but by the increasing number of consumers [population].*

> *Nothing in political economy can be of more importance than to ascertain the law of this increase of production.* (p153)

Jevons reported many statistics on the increase of both per capita wealth and population since the 18th century (ppvi, 196–200 and 457). He moreover both extolled and feared for Britain's prosperity and greatness: the 'Age of Coal' enabled:

> *A multiplying population, with a constant void for it to fill; a growing revenue, with lessened taxation; accumulating capital, with rising profits and interest.* This is a union of happy conditions which hardly any country before enjoyed, and which no country can long expect to enjoy. ... *It is the very happiness of civilization.* ... *[Without coal] we must ... sink down into poverty [and]* begin a retrograde career. (pp2, 231, 11, 201 and 454–460; emphasis original)

He quotes Baron Liebig that civilization 'is *the economy of power*' (pp142 and 156). And since for Jevons the greater economy of coal increased not only affluence but its quicker exhaustion, *'We have to make the momentous choice between brief but true greatness and longer continued mediocrity'* (p460). The discussion today likewise contains the political hope that energy efficiency is the key to both happy prosperity and sparing natural resources. Now, as then, we should not ignore our normative assumptions.

That the correlation between consumption and efficiency reflected causality was, to be sure, denied by no one. Clarity has reigned from Petty onward on the point that *quantities* of land, labour or capital do not suffice to explain the size of the wealth of a nation.[46] The causal factor for greater wealth, produce, riches, returns and surpluses was higher productive powers of land and labour, often aided by invention and machines.[47] Mill even asserted that 'improvements ... by the very fact of their deserving that title, produce an increase of return' (p93) and elsewhere *equated* 'the magnitude of the produce' with 'the productive power of labour' (p413). Today also this seems self-evident.

Even for Malthus, despite his observation that we could always choose to really save through indolence or direct non-consumption, the doctrine was that 'the increased powers of labour would *naturally* produce an increased supply of commodities' (p63, emphasis added). Say said that although lower input and greater output are mathematically 'the same thing', both are 'sure to be followed by an enlargement of the product'; for both producers and consumers 'every thing saved is so much gain' (pp301 and 357). It was Rae who, while concurring with the standard causal chain from increased capital through increased division of labour to increased wealth, shifted the emphasis from organizational to technological efficiency: it is 'the intention of the inventive faculty', which creates and improves instruments, to increase 'necessaries, conveniences or superfluities' and make 'larger returns', 'supplies', 'absolute capital and stock', 'revenue' and 'supply for future wants' (pp67 and 258–260; see also Brewer, 1991). For him the 'effective desire of accumulation' was necessary but not sufficient for the 'increase of stock and capital', which also required 'augmentation', that part of growth occurring 'through the operation of the principle of invention' (pp205–209 and 264 and Chapters VI and VII; see also Malthus, p339). And since invention results in higher efficiency, a causal arrow goes from efficiency to 'larger provision ... made for the future wants of the whole society' (p165). Since instrument formation means cost and 'sacrifice' in the present, without 'some future greater good ... the instrument ... will not be formed', yet this results only from greater efficiency (pp19, 110–118 and 171).[48]

If pressed, no classical economist would have claimed that he was describing mere correlation rather than causality. And since all wealth requires material inputs, in any description of the 19th-century economy, rebound is certain and a low rebound out of the question. Without efficiency increases and given only certain *quantities* of material resources and labour, not much more in the way of food or any other goods can come into existence; and unless we enjoy these (labour) efficiency increases wholly and exclusively as the less work and more leisure that they enable, there is *some* consumption that wouldn't be there without the 'improvements'. And this consumption depends on labour and material inputs. Until Jevons, however, the doctrine did not attest backfire. Before surveying classical views on the magnitudes of this new consumption of goods and services, and their inputs, let us relate their descriptions more closely to today's debate by introducing the term *prices* and the *price falls* that result when a good is produced with lower input.

PRICE FALLS

In 1815, Ricardo wrote to James Mill, 'I know I shall soon be stopped by the word price, and then I must apply to you for advice and assistance' (Sraffa, 1951, pxiv). And no classical economist failed to warn of conflating money and wealth, with the term 'value' leading an ambiguous life between the two.[49] But being economists, our previous writers could not avoid monetary terminology altogether. While prices can be physically expressed as exchange value in terms of other commodities, the monetary metric is convenient. Thus all of them presaged the point made by Khazzoom in re-opening the debate over the Jevons Paradox that efficiency increases have a 'price content' (1980, p22). In Smith's analysis, for instance:

> It is the natural effect of improvement ... to diminish gradually the
> real price[50] of almost all manufactures. ... In consequence of better
> machinery, of greater dexterity, and of a more proper distribution of

work ... a much smaller quantity of labour becomes requisite for executing any particular piece of work; and though, in consequence of the flourishing circumstances of the society, the real price of labour should rise very considerably, yet the great diminution of the quantity will generally much more than compensate the greatest rise which can happen in the price. (I.xi.o.1 and I.viii.57; see also Jones, p238; Marx, p379)

Although Smith here succumbs to the tendency to exogenize a vague 'flourishing circumstances of the society', the point is well made that because improvement more than compensates rising input prices, output prices fall. He then considers rising and falling prices of 'rude material' and metal inputs together with a comparison of output prices over three centuries (I.xi.o.2–13; see also Barnett and Morse, 1963).

In Malthus's formulation, 'We all allow that when the cost of production diminishes, a fall of price is almost universally the consequence' (p60; see also pp87–88 and 145).[51] Favourite empirical examples were cottons in general and stockings in particular.[52] Printed goods likewise had experienced a palpable, undeniable 'reduction in price' per copy (Say, pp88 and 302). Rae liked the example of more efficiently produced, cheaper bread (p259; Mill, p181), while Mill liked Say's 'still stronger example' of playing cards (p123). Babbage's example of riveted tanks showed an extreme price fall (p100). Malthus even distinguishes between 'a fall of price necessary ... to prevent a constant excess of supply contingent upon a diminution in the costs of production' and one following 'an increased supply of commodities' albeit itself due to 'the increased powers of labour' (pp56–57 and 63).[53]

The necessity of this step from efficiency increase to price fall – and then on to consumption increase – lies in producer behaviour. '[C]ompetition of producers brings the price of the product gradually to a level with the charges of production', wiping out temporarily high profits (Say, pp93 and 395). Of course, patents must first run out or secrets be divulged, but eventually 'The grinding of corn is probably not more profitable to the miller now than formerly; but it costs infinitely less to the consumer' (Say, p89). For Rae, still in monetary terms, each of:

the vendors of a commodity wishes to sell as much as possible, and as he can do so most readily by underselling his neighbours, the price gradually falls under a free competition, until the dealers in it receive only the profits that the effective desire of accumulation, and the progress of improvement in the society measures out to them. (p307)[54]

Mill also pointed to producers' 'power of permanently underselling' which can 'only ... be derived from an increased effectiveness of labour' (p133 and 495). Jevons relied on this argument from profitability (pp8, 141 and 156) and names the 'series of inventions' by Bessemer, Gilchrist and Thomas as 'modes of economy which, in reducing the cost of a most valuable material, lead to an indefinite demand' (p390).

Rae solves the profits 'paradox' thus: 'Now I apprehend that high profits springing from improvement can never lessen the sale of goods either at home or abroad, for they do not occasion a rise in their price, but rather a fall in it.' (p263) Domar's later version is that 'a rapid growth of [Kendrick's] Index [total factor productivity] in any industry reduces the prices of its output, and thus stimulates sales' (1962, p605).[55] Malthus once chastises Ricardo for ignoring this point and in effect assuming that profits stayed high – 'at *cent* per cent' (Malthus, p291). Moreover, whatever the profit-maximizing price policy of a monopolist is, even monopoly profits get spent because, in Say's terms, producers are also consumers (p89; see also Smith, I.xi.o.4; Ricardo, pp386–387 and 392–394). This fact casts doubt on today's view that rebound is low in sectors where 'market failures' are high (Grubb, 1990b, pp783–785; 4CMR, 2006, pp5 and 14).[56]

The classical axiom is that prices of output are the sum of the prices of inputs or charges of production (Ricardo, p397). Say talks of 'a real fall of price, or *in other words*, a reduction in the price paid to productive exertion' (p303, emphasis added).[57] Output and input prices are exactly proportional. Supply costs fall, prices fall, effective demand rises, number of units sold rises; these are today's 'price and income effects' of efficiency increase.[58] Rebound is then a function of this new quantity sold (Q) after deducting another quantity no longer sold (Q_s) of units, if any, *for which* the newly more efficiently produced item is substituted.

As for price elasticity of demand, Malthus writes that 'the increase in the whole value of cotton products, since the introduction of the improved machinery, is known to be prodigious', offering the empirical evidence of 'the greatly increased population of Manchester, Glasgow, and the other towns where the cotton manufactures have flourished' (p192; see also pp281–282; Rae, p292) Say observed the same for 'Amiens, Rheims, Beauvais ... Rouen and all Normandy', where there had first been 'loud remonstrances' over the annihilation of local industry, and gives further examples of 'prodigious' price falls (pp147–148 and 300–304); he then can't resist imagining prices falling to zero, which would at once be 'the very *acme* of wealth' and the death of political economy as a science (p304). Finally, Mill makes the empirical macroeconomic claim of falling prices over two centuries, 'accelerated by the mechanical inventions of the last seventy or eighty years' (p182). All these economists were describing, via price falls, a very high 'efficiency elasticity of demand' (Sorrell and Dimitropoulos, 2006, p7). But demand for what? For the newly cheaper good? For everything, as described in the next section? For our topic of interest, material and labour inputs?

But as long as we are thinking in monetary terms, what happens to the *total* amount of money paid for the goods now cheaper *per unit*? This is the new price per unit times the new quantity (P × Q) as opposed to the new quantity physically measured (Q) and was termed by Say *'le montant total'* or sum total (p450). He gives a descriptive example of (direct) backfire in the 'art of printing':

> *By this expeditious method of multiplying the copies of a literary work, each copy costs but a twentieth part of what was before paid for manuscript; an equal intensity of total demand would, therefore, take off only twenty times the number of copies; probably it is within the mark to say that a hundred times as many are now consumed. So that, where there was formerly one copy only of the value of 12 dollars ... there are now a hundred copies, the aggregate value of which is 60 dollars, though that of each single copy be reduced to 1/20.* (p302; see also Rae, pp216 and 249–250)

Taking price and costs as equal and substituting 'labour time' or 'material amount' for 'dollars', we can estimate input consumption. Substituting 12 hours of labour for 12 dollars, if the price elasticity of demand is in a ratio of 20:100, in the end 60 hours of labour are demanded and labour input demanded is higher than it *would have been* without the efficiency increase. Say could analogously describe 'direct backfire' regarding energy efficiency today.

Still referring only to manufactured goods made cheaper, rather than the whole economy, Malthus writes that by means of:

> *the introduction of improved machinery, and a more judicious division of labour in manufactures ... not only the quantity of manufactures is very greatly increased, but ... the value [price, cost] of the whole mass [P × Q] is augmented, from the great extension of the demand for them both abroad and at home, occasioned by their cheapness. ... The reader will be fully aware that a great fall in the price of particular commodities ... is perfectly compatible with a continued and great increase, not only in the exchangeable value of the whole produce of the country, but even in the exchangeable value of the whole produce of these particular articles themselves.* (pp135 and 314)[59]

While Khazzoom's demonstration of rebound assumed *any* positive price elasticity of demand (1980, p22), Malthus describes a very high elasticity. The point, in Say's words, is that 'every real reduction of price, instead of reducing the nominal value of produce raised, in point of fact augments it' (p303). P × Q for product or sector X increases following productivity-induced price falls. Following Say that work is done by nature (for example fossil fuels) as well as human beings, in other words it commits 'productive exertion' (pp40, 63, 74–75, 90 and 245 note; Rae, pp246 and 256–258); we have, for any X, P_{Labour} and $P_{Material}$ both as costs and prices. $Q \times P_{Material}$ after an efficiency increase is compared with that before, but where Q rises by any amount, direct rebound is proved. The relative degrees of growth of Q and P_{Material} determine the size of this direct rebound.

But what happens in sectors not affected by productivity increases? Or how can the 'value of the whole mass' (economy-wide) increase unless *money supply* increases? If it doesn't, *less* demand would have to accrue to goods that did not enjoy a productivity increase.[60] And monetarily, the consumer's gain is perhaps equal to the producer's loss. Monetary analysis also entails identifying cases where substitution of the newly cheaper good for another good occurs, then measuring both the price and the substitution elasticities. Should rebound research discard the veil of money and deal only with Q to joules, with each unit Q measured physically – rather than compare ratios of P × Q to joules before and after an efficiency shock, as with the concept of energy intensity of a unit of GDP?

Mill's heroic attempt to sort out the concepts of price, use value and exchange value and their application to particular goods as opposed to the whole mass (pp455–459) relegates 'price' to goods' relationship to money and 'exchange value' to an economic discourse dispensing with 'money', namely to 'the command which [a good's] possession gives over purchasable commodities in general' (p457).[61] He also made the point that 'if inventions and improvements in production were made in all commodities, and all in the same degree, there would be no alteration in [relative exchange] values' (p710). But Say (pp303–305) and Malthus (p135), even when using the term 'exchange value', were talking not sectorally of the 'values' or prices of things relative to each other but of the 'whole mass', conceivably tradable for other things in other countries. Criticizing his predecessors in all but name, Mill concludes that 'All commodities may rise in their money price. But there cannot be a general rise of values' (p459).

Mill has a point. If, as Malthus somewhat circularly said, 'exchangeable value is the relation of one object to some other or others in exchange' (p51), then the concept of exchange is of no use in analysing the *growth* of wealth. And to the extent that prices are an abstract proxy for millions of exchange values, monetary concepts are likewise perhaps inapplicable. In Malthus's words:

> When it is said that the exchangeable value of a commodity is determined by its power of purchasing other goods, it may most reasonably be asked what goods? It would be absolutely impossible to apply all goods as a measure. (p97 note)

This does not prevent Malthus elsewhere from talking of 'the increase in the exchangeable value of the whole produce estimated in labour' (p192) and even of the value of money expressed in labour (p144 note). And after listing shortcomings of *any* metric of value which remind one of today's criticisms of GDP, he opines that we can't do without one, if only to compare the total products of different economies (pp247–248 and 255–256). Such difficulties in integrating concepts of exchange and price with the 'value of the whole mass' arise in Rae's struggle with the paradox that a limited amount of exchange value in terms of prices coexists with greater wealth, and he concludes that the relevant magnitude was the *physical* increase in 'absolute capital and stock' (pp259–260).[62]

Whatever happens economy-wide, price falls and underselling of more energy-efficient goods raises their relative attractiveness. Jevons used the common classical phrase that coal 'commands' iron and steam (p2; Martinez-Alier, 1987, p161); whatever is more cheaply or powerfully commanded – products requiring iron and steam – enjoy higher demand. If I can commute to work by bicycle, bus, horse, car or on foot, more efficient motors give the car the edge. This implies high economy-wide or total rebound and even backfire even if economy-wide Q or P × Q does *not* increase – a pure 'substitution' effect distinct from income effects and the derived categories of 'direct' and 'indirect' rebound.

The purely physical perspective shows us that the actual amount of coal or oil for a steam engine, car or light bulb over its 'lifetime' drops, enabling us to ask after the price or exchange-value effects on the *inputs themselves* rather than the outputs such as a pair of stockings: the initially lower demand at constant output lowers the input price, in turn raising demand for it relative to all else. Combining this aspect with the income effects discussed in the next two sections, Burniaux et al, for instance, write:

> *There is a link between technical progress, output prices and real income. ... the rise in energy productivity tends to lower the relative price of energy, thereby generating a substitution effect from non-energy [sic] towards energy goods. In the aggregate the increase in autonomous energy efficiency also generates a real income gain that leads to higher consumption of both energy and non-energy goods. The net result is that emissions do not decrease in the same proportion as the AEE [autonomous energy efficiency] increase because the energy conservation effect is partly compensated by the relative price and income effects.* (1995, p246; see also Hearn, p99)

The size of this input-price-determined rebound effect depends also on the price elasticity of supply, for example of petroleum. At any rate, empirical work must analyse energy prices as well as efficiency change and change in the consumption of 'outputs'.[63]

SOCIETAL INCOME EFFECT

Smith's 'invisible hand' is not all that invisible but a name for the mechanism starting with efficiency increase, in other words with dexterity, division of labour, trade and machines 'directing ... industry in such a manner as its produce may be of the greatest value' – a 'greatest value' variously called 'wealth', the 'annual revenue of the society', its 'power of purchasing', or 'the exchangeable value of the whole annual produce of its industry' (IV.ii.4 and 9, I.iv.13, I.vi.17 and II.ii.21). This revenue or purchasing power – concepts closer to consumption than to production – was divided between labour/wages, capital/profits and land/rents, raising the allocative question which for Ricardo was the defining *explicandum* of political economy (pp5 and 347).[64] While the others likewise devoted much effort to this issue,[65] their main concern was the question of scale, or the size and growth of production and consumption (Daly, 1992). Malthus even castigates Ricardo by name, writing that 'to estimate rent and wages by the *proportion* which they bear to the whole produce must, in an inquiry into the nature and causes of the wealth of nations, lead to perpetual confusion and error' (Malthus, p164). More politely, Say remarks of landowners and capitalists, 'The world at large may be content to comprehend, without taking the trouble of measuring, their respective shares in the production of wealth' (p74 note).

Rae conceptualized the crucial distinction with the terms 'acquisition' and 'augmentation' (sometimes 'accumulation'); the former is a mere shift of wealth from one person, group or nation to another, the latter a rise of the total (or per capita average) amount of produce (pp11–12, 24, 264 and 307; Say, p85; Malthus, p35; Mill, p62). Following Say (pp70 and 117–118) he names this 'creating wealth', claiming that 'the ends which individuals and nations pursue are different. The object of the one is to acquire, of the other to create' (Rae, p15). 'As individuals seem generally to grow rich by grasping a larger and larger portion of the wealth already in existence, nations do so by the production of wealth that did not previously exist' (p12). Not Smith's invisible hand, but the state or 'community' must promote and encourage 'progress of art', the 'discovery of new arts' and the 'discovery of improvements in the arts already practiced in the country' (pp15 and 12).

The clearest description of the augmentation of societal income is Say's:

> ... *the aggregate utility will be augmented; the quantum of products procurable for the same [total] price will be enlarged. ... But whence is derived this accession of enjoyment, this larger supply of wealth, that nobody pays for? From the increased command acquired by human intelligence over the productive powers and agents presented gratuitously by nature. A power has been rendered available for human purposes, that had before been not known, or not directed to any human object ... or one before known and available is directed with superior skill and effect, as in the case of every improvement in mechanism, whereby human or animal power is assisted or expanded.* (p299)

Say sharpened this concept of greater wealth that nobody pays for by expanding his system boundary to include the whole world, describing sales between nations as mere acquisitions in Rae's sense and insisting that 'the general stock of wealth, existing in the world ... can only be enlarged by the production of some *new utility*' (p305, emphasis added; see also p318). Malthus later described this shift of the societal supply curve caused by lower costs of production as a change in the 'conditions of supply ... advantageous to the consumer' (1825, p303).[66] Mill as well identified this rise in 'general purchasing power', caused for instance by 'an invention ... made in machinery, by which broadcloth could be woven at half the former cost'; for him, simply, 'All ... improvements make the labourers better off with the same money wages' (pp457–458 and 751).

'Wealth, that nobody pays for'? Is there a free lunch after all? (Jones, pp288–289) Evidently yes, once inventors, research and development, and embodied inputs are deducted as costs. The point is that the source of this lunch is efficiency. This productiveness inheres either in nature, as with increased dexterity or education of humans and the substitution of naturally better materials, or in their ways of organizing themselves and their materials by 'forming' or 'transforming' matter for utility (Cantillon, p2; Say, pp62, 65 and 387; McCulloch, p61; Rae, pp81–83).[67] Virgin land, virgin mines and population growth can bring greater output for constant input per unit, but efficiency brings this result even when the limits of these things are reached, or closely approached.

Once Say had fingered this win–win process he defended it with sarcasm, as against Galiani and Forbonnais, whose idea that one's gain must be another's loss underpinned the 'systems of all the short-sighted merchants' (Say, pp16, 31 note and 70). More didactically and again reflecting the struggle with the term 'value' he wrote:

> *If different commodities have fallen in different ratios ... they must have varied in relative value to each other. ... There is this difference between a real and a relative variation of price [*valeur*]: that the former is a change of value, arising from an alteration of the charges of production; the latter, a change, arising from an alteration of the ratio of value of one particular commodity to other commodities. Real variations are beneficial to buyers, without injury to sellers, and* vice versa, *but in relative ones, what is gained by the seller is lost by the purchaser, and* vice versa. (Say, p304; see also Mill, 457–458)

His summary:

> *In commercial, as well as manufacturing industry, the discovery of a more economical or more expeditious process, the more skilful employment of natural agents, the substitution, for instance, of a canal in place of a road, or the removal of a difficulty interposed by nature or by human institutions, reduces the cost of production, and procures*

a gain to the consumer, without any consequent loss to the producer, who can lower his price without prejudice to himself, because his own outlay and advance are likewise reduced. (Say, p101; see also pp89 and 301)

Say later offers a numerical example expressing purchasing power in terms of 'the quantity of his own particular product' instead of money: once stockings are made cheaper, a sugar tradesman can get the same number of stockings as before for less sugar (p300). He then assumes simultaneous price falls of sugar and stockings, asking whether we are now:

authorized to infer that this fall is a positive fall and has no reference or relation to the prices of commodities to one another? That commodities in general may fall at one and the same time, some more, some less, and yet that the diminution of price may be no loss to anybody? (pp300–301)

McCulloch argued against the claim that consumers' gains might be balanced by producers' losses, and in his own jibe at Ricardo also saw win–win cases where 'profits ... would have risen, without their rise having been occasioned by a fall of wages' (p372). Distribution is here not the issue. Malthus also empirically attests rising profits and, moreover, lest anyone fear slacking demand, capitalists' rising expenditures 'in objects of luxury, enjoyment and liberality' (p293). While arguing that labour efficiency causes unemployment, Sismondi had ignored this point that demand for labour originates from profits as well (Sismondi, vol 2, pp322–324 and 335). Jevons later added that even when profits through competition fell to their minimum, there is a net gain to society (1871, p254).

The possibility, however, that suppliers' profits as a total amount of purchasing power could fall seemed real. Charles Babbage 'strongly pressed upon the attention' of the manufacturer to very carefully 'ascertain how many additional customers he will acquire by a given reduction in the price of the article he makes' lest profits turn to losses, adding that falling prices would force firms to make further efficiency gains (pp98–99; see also Say, p87).[68] Old goods produced more expensively, for instance, must be sold at a loss (albeit a gain for the consumer) (Say, pp305 and 390; Ricardo, p274; Malthus, p282).[69] The profits of the producers of material inputs – for example of energy or mining companies – could also fall since they experience at least initially lower demand and must lower prices; however, the rebound caused by lower input prices in the longer term restores profits.

Smith was describing this economy-wide income effect of newly enabled, costless prosperity by writing for instance that 'all things would have become cheaper in reality'; 'improvements in mechanicks ... are always regarded as advantageous to every society'; the surpluses of 'the country', division of labour and trade with 'the town' raise the revenue of *both* (I.viii.4, II.ii.7, III.i, IV.vii.c.88, IV.ix.51, IV.ix; Mill, pp119–122). For Rae 'all instruments at the period of their exhaustion return more than the cost of their formation' (p118) and 'good bread ... produced ... with half the labour and fuel ... would not benefit the bakers exclusively, but would be felt equally over the whole society' (p259). Efficiency is like corn – one seed yields 100 seeds. Jevons likewise later wrote that profits falling to their minimum means that everything is cheaper, and that 'either the labourers themselves, or the public generally as consumers, gather all the *excess of advantage*' (1871, pp254 and 257, emphasis added). Finally, Mill quoted Rae's description of the contrasting 'stationary state' society of China (Mill, pp168–169) and referred to the free increase of wealth caused by 'improvement' as an 'increased means of enjoyment' (Mill, p724).

If we now make the attempt to approach rebound while ignoring prices, as suggested in the last section, we can for instance assume that before an efficiency increase production is 10X at 10 joules/X equalling 100 joules of input. If afterwards there are 12X at 9 joules/X, this equals 108 joules of input, in other words backfire. Our writers often claimed that this is the normal case: we can produce not only 10 per cent more X if efficiency increases 10 per cent, but 20 per cent. Is this something coming from nothing? It is easy to accept that 11X are produced, using 99 joules of input, in other words rebound of 100 per cent. But whence the 12th X? The source can only lie with increased purchasing power due to X's price fall, with purchasing power seen as an income effect, or taken away from rival factors of production like labour, or due to a price fall *of the input joules*.

One argument for the possibility of backfire thus does not depend on the concepts of societal income effect or even growth of total output: if a given factor of production becomes more powerful, to use the classical term, demand for that factor will increase relative to rival factors of production whose productiveness remains the same (Marx, p354; Brookes, 1990 and 2000; Saunders, 1992 and 2000a). Brookes writes:

> *The market for more productive fuel is greater than for less productive fuel, or alternatively ... for a resource to find itself in a world of more efficient use is for it to enjoy a reduction in its implicit price with the obvious implications for demand [for fuel].* (2000, p355)

Jevons similarly concluded his chapter 'Of the economy of fuel' by asserting necessary rises in both input and output consumption:

> *And if economy in the past has been the main source of our progress and growing consumption of coal, the same effect will follow from the same cause in the future. Economy multiplies the value and efficiency of our chief material; it indefinitely increases our wealth and means of subsistence, and leads to an extension of our*

> *population, works and commerce, which is gratifying in the present, but must lead to an earlier end. Economical inventions are what I should look forward to as likely to continue our rate of increasing consumption.* (p156)[70]

Again, if we interpret the societal income effect monetarily we encounter the paradox that a consumer with a new park of efficient appliances pays less to the electricity supplier, lowering his income, purchasing power or consumption. Where a high price elasticity of demand is claimed (for example Say, p302, or Malthus, p192), we could encounter a book-keeping quantity 'that nobody pays for': if before an efficiency event 36 units are sold at £2 each, P × Q = £72, and where price elasticity of demand is 1, 72 × £1 also = £72. If price elasticity of demand is 2, then 144 units sold yields £144. Whence the additional £72? If withdrawn from sectors previously favoured, we must deduct this from rebound. Again, it seems clearer to simply realize that *more output* is here at the same cost. If societal purchasing power is £1,000,000 and newly more efficiently produced things are now £1000 cheaper, we have a monetary hole that gets filled up with material goods.

HIGH REBOUND

One conclusion till now is that efficiency-induced consumption of output, entailing as it does *some* input, proves rebound. Before looking more closely at classical descriptions of high rebound, some taxonomy is useful.[71] Increased society-wide purchasing power results from the increased efficiency of producing an average unit of a good of type X, as opposed to Y, representing all other goods. At this moment, as Malthus said, 'there must be a considerable class of persons who have both the will and power to consume more material wealth than they produce' (p319). This new demand can be:

1   for additional X by consumer A, a previous consumer of X;
2   for some Y by consumer A;
3   for additional X by a new 'marginal' consumer B;
4   for some Y by consumer B, who after consuming some X retains some 'consumer surplus'; and
5   for leisure – in the extreme, all consumers choose to lower their purchasing power to the full extent of engineering savings.

Aside from these variations of the income effect, a more efficient production factor is substituted for another one – a 'substitution' effect.

The first case is called 'direct rebound', today's workhorse example being that if my new car uses less petrol per kilometre, my existing purchasing power allows me to drive more kilometres; this is Khazzoom's 'own' price elasticity of demand (1980, p22). The total cost of the car, including its use, has dropped, freeing income.

The second case is in Malthus's words 'distinct from' the first and pertains when 'the commodity to which machinery is applied is not of such a nature that its consumption can extend with its cheapness' but 'there would be a portion of revenue set free for the purchase of fresh commodities' (pp282–283). Given higher purchasing power, when the price elasticity of demand for the newly cheaper good is low, *indirect rebound* results (even with high efficiency elasticity of price). In unfairly claiming that Malthus missed this point, McCulloch offers a clear description of it:

> *Suppose the price of cottons were reduced in the proportion of ten to one; if the demand for them could not be extended, it is certainly true that nine tenths of the capital and labourers engaged in the cotton manufacture would be thrown out of* that *employment. But it is equally certain that there would be a proportional extension of the demand for the produce of* other *branches of industry. It must be remembered that the means by which the purchasers of cottons formerly paid for those that were high priced could not be diminished by the facility of their production being increased and their price reduced. They would still have the* same capital *to employ, and the* same revenue *to expend.* (pp177–178 and 188, emphasis added)

The indirect rebound of the second and fourth categories above is likewise in Say's remark that:

> *A new machine supplants a portion of human labour, but does not diminish the amount of the product; if it did, it would be absurd to adopt it. When water-carriers are relieved in the supply of a city by any kind of hydraulic engine, the inhabitants are equally well supplied with water. The revenue [purchasing power] of the district is at least as great, but it takes a different direction. … [I]nferior charges of its production [mean that] the revenue of the consumers is benefited.* (pp86–87)

Say's translator Prinsep is explicit: 'our revenues are enlarged by lower costs of production of X, and we are free 'to employ them upon some other object [types 2 and 4], or upon an enlarged production of the same object [types 1 and 3]' (p296 note). Ricardo likewise, quoting Smith's attestation of unlimited desires for all but food, brings the example where 'improved machinery, with the employment of the same quantity of labour' quadruples 'the quantity of stockings

[but] the demand for stockings were only doubled', leading to 'the production of some other commodity' (p387). In Malthus's version:

> ... *though the wills and means of the old purchasers might remain undiminished, yet as the commodity could be obtained without the expression of the same intensity of demand as before, this demand would of course not then show itself.* (p55)

Based on this consumer surplus, demand could and would show itself elsewhere.

In the classification above good Y could also be a *new good*, in other words one not existing at the time of the efficiency increase but whose supply and consumption depends on that efficiency increase. Examples are legion – railways following better steam engines and cheaper steel, or emails following the more efficient use of electricity in data transmission. Transportation, milling, printing and glass-making all count for Rae as consumption areas opened up by efficiency (pp116–117, 245–250 and 291–292) while Hearn presaged Jevons's emphasis on new uses and products in observing that:

> *In many districts the price of coal has been reduced from thirty to forty per cent; and the purposes to which it has been applied have consequently been largely increased.* (p274)

Jevons repeated this general point (pp141–142 and 197) and named new uses of coal in metallurgy and transportation (see footnote 23). Martinez-Alier points out that instead of substituting for coal, electricity increased demand for it (1987, p88; Jevons, p181). Sanne draws the exact parallel with new applications of electricity as it becomes cheaper due to increased efficiency of coal-fired plants (2000, p489).

Jevons called this new consumption 'the reaction and mutual dependence of the arts' as when Darby's powerful-blast smelting oven required the substitution of coal for water (pp372 and 385). And the fundamental phenomenon of productivity's opening up *new* markets had been sketched early on by Smith (I.xi.c.36) and filled out somewhat by Say (pp89–90) and Rae (pp245, 247 and 253). But granted that 'many of the more important substitutions are due to coal' (Jevons, p134), what are the net effects? Coal's efficiency meant that fewer horses and oats were consumed due to railways, just as today efficiencies of electricity production and use mean perhaps that fewer paper letters are sent due to email. Again, how much of this new consumption should be booked under rebound is hard or impossible to decide, and while today it is implicitly subsumed under 'economy-wide' rather than either direct or indirect rebound, it is *ignored* by all rebound studies. Fresh study, for instance of Babbage, von Liebig (1851), Cipolla (1962), Rosenberg (1982 and 1994), Clapp (1994) or Sieferle (2001), is warranted.

As for the direct rebound of the third case, where marginal consumer B purchases X, all writers observed that the efficiency-induced cheapening of X enables marginal consumers to buy it; how much of this demand is truly new, in other words not shifted from Y, is an open question. Say writes:

> *Suppose that ... knit-waistcoats of wool [cost] 2 dollars each ... those who should have but a dollar and a half left must ... go without. If the same article could be produced at one dollar and a half, these latter also might all be provided and become customers; and the consumption would be still further extended if they should be produced at one dollar only. In this manner, products formerly within the reach of the rich alone have been made accessible to almost every class of society, as in the case of stockings.* (p288)

Malthus echoes Say, talking of:

> *such an extension of the demand for the commodity, by its being brought within the power of a much greater number of purchasers, that the value of the whole mass of goods made by the new machinery greatly exceeds their former value'.* (Malthus, p281; see also p314)

In terms of I = PAT (that is, Impact = Population × Affluence × Technology), $(P \times A)_{after} > (P \times A)_{before}$. Sismondi reminded these economists, however, that since the laid-off workers have no more purchasing power the market extension is inhibited (vol 2, pp251, 316–317 and 326–327). We can moreover ask Say and Malthus what the marginal consumer had done with his one dollar and a half before the price of the waistcoat fell from 2 dollars. Whatever would have been consumed without the cheapening of the waistcoat is no longer consumed, constituting to some degree a win–lose situation after all.

Also part of 'indirect' rebound is the fourth category where a marginal consumer's demand for X evidences some consumer surplus, leaving some purchasing power for Y. Taken together the four categories equal total rebound or the societal income effect. Today all rebound researchers acknowledge the difficulty of tracing these effects from direct rebound through indirect rebound to what really matters, namely total or economy-wide rebound. Wirl notes that excluding 'marginal consumers' gets around the 'conservation [or] energy paradox' but yields an underestimation of rebound (1997, pp19–32, 36 and 112). Roy believes that there is 'a whole range of behavioural responses of the end-users that follow any technical efficiency improvement all of which may, however, not be traced empirically' (2000, p433).[72] What then are we to make of Allan et al's assertion that 'rebound is an empirical issue. ... It is simply not possible to determine the degree of rebound and backfire from theoretical considerations alone' (2006, pp21–22; see also pp3 and 10)?

Malthus already saw this. Assuming, he said, that latent demand in the affected sector was low:

> *To what extent the spare capital and labour thrown out of employment in one district would have enriched others, it is impossible to say; and on this subject any assertion may be made, as we cannot be set right by an appeal to facts.* (p286)[73]

It is likewise doubtful whether we today have the data – the 'facts' – necessary for demonstrating that a given increase in one sector constitutes indirect rebound from efficiency in another sector. Direct rebound is apparently more easily estimated. Some of these sectoral studies calculate high direct or even total rebound (Dahmus and Gutowski, 2005; Allan et al, 2006; Herring, 2006; Fouquet and Pearson, 2006) while some, implicitly or explicitly offering support to the environmental efficiency strategy, show total rebound as low as 26 per cent and thus real energy savings (4CMR, 2006, pp6, 9 and 66).[74] Other studies attest low rebound while limiting themselves to direct rebound and moreover equivocating between direct and total rebound (Greening et al, 2000; Berkhout et al, 2000).

The fifth category, wherein leisure is chosen, is crucial: rebound *can* be zero if price elasticity of demand is vertical. As shown in the next section, only Malthus gave weight to this possible reaction, the others agreeing with Rae that 'improvement [is] absorbed by vanity' (pp289–290) or with Jevons that children will continue doing as their elders did (p199). That is, humankind finds itself in a condition far from saturation. To attest rebound is merely to assert that short of total consumer satiation, theoretical input savings are never fully realized, whereas backfire depends upon a strong low-saturation premise. The sixth category of 'substitution' effects, which includes the effects of a fall in the *input* price relative to other prices, received little explicit attention in Jevons and the classical literature.

The classical input metric was not always labour, land area and mines. Mill once observes that 'the tendency of improvements in production is always to economize, never to increase, the expenditure of seed or material for a given produce' (p99). And renewable energy resources concern him in his analysis of the invention of – *nomen est omen* – windmills and watermills (p28). Rae was more explicit:

> *Every society possesses a certain amount of materials capable of being converted into instruments. The surface of its territory, the various minerals lying below the surface, its natural forests, its waters ... are all to be regarded as materials, which, through the agency of the labour of its members, may be converted into instruments. The extent of the power, which the inhabitants of any state may possess, to convert into*

instruments ... is however variable and increases ... as their knowledge of the properties of these materials and of the events [products], which in consequence of them, they are capable of bringing to pass, increases. [K]nowledge ... gives ... the power of constructing a much greater number of instruments out of the same materials. (p99)

This leads to Rae's long chapter on invention, which always serves efficiency either by changing 'materials' or applying given 'materials' to new arts (pp224–229, 242–249 and 258–259). In Smith (I.xi.o.12), Say (pp89–90) and Rae (pp242–244) the insight is that without inventions, water and wind are not used at all, but that once the right equipment is available, the energy is used more and more. The bridge from invention to efficiency is established by Jevons's closely related, ironic observation on the difference between Savery's coal-burning steam engine and those of Newcomen and Watt: Savery's *'consumed no coal, because its rate of consumption was too high'* (p143). Once invention has occurred, the consumption of an input is positively proportional to the efficiency of its use – yielding rebound for sure but not necessarily backfire.

SURPLUS AND INDOLENCE

Malthus threw a monkey wrench into the mechanism of output growth described by Smith, Say, Ricardo and himself:

It has been supposed that, if a certain number of farmers and a certain number of manufacturers had been exchanging their surplus food and clothing with each other, and their powers of production were suddenly so increased that both parties could, with the same labour, produce luxuries in addition to what they had before obtained, there could be no sort of difficulty with regard to demand. ... But in this intercourse of mutual gratifications, two things are taken for granted, which are the very points in dispute. It is taken for granted that luxuries are always preferred to indolence, and that an adequate proportion of the profits of each party is consumed as revenue. The effect of a preference of indolence to luxuries would evidently be to occasion a want of demand for the returns of the increased powers of production supposed, and to throw labourers out of employment. (Malthus, p258; see also p9)

Greater consumption following increased efficiency is not necessary but only what 'almost always happens' (p170). What if, he asks, 'after the necessaries of life were obtained, the workman should consider indolence as a greater luxury than those which he was likely to procure by further labour' (p268)? 'The peasant, who might be induced to labour an additional number of hours for tea or tobacco, might prefer indolence to a new coat' (p283). In richer societies, likewise, it could be that the 'habits and tastes of the society prevent [an] increased consumption' and 'the demand for material luxuries and conveniences would very soon abate' (pp191 and 288; see also Mill, p105) – the vision of today's sufficiency strategy (Alcott, 2007).[75] Even for poorer societies like that of North American Indians, whose 'proverbial indolence' he attests, the rule is that 'to civilize a savage, he must be inspired with new wants and desires' (Malthus, pp103–104).

Malthus's population essay already notes these limits to demand for produced goods (1798, pp95 and 120). However, he knows that the 'laws of nature have provided for the leisure or personal services of a certain portion of society', and that the tastes and habits of this leisure class (Veblen, 1899), perhaps due to exposure to items of foreign trade, can sustain a good deal of luxury consumption (Malthus, pp317 and 284). The issue here is not 'Say's Law' – that overproduction is only temporary – but human psychology. Jevons explicitly maintained that we cannot count on consumption or reproduction desires subsiding, and even claims this to be 'the gist of the subject' (p194). He knew that his argument that fuel's very economy was part of the problem needed assumptions about desires, saturation and demand elasticities: the 'natural laws [of growth] which govern ... consumption' (pp25 and 275) must be firmly assumed in our models of energy use. To be sure, he frames the classical view both of population increase and the desire for greater and greater material wealth in the conditional:

> *If our parents doubled their income, or doubled the use of iron, or doubled the agricultural produce of the country, then so ought we, unless we are changed either in character or circumstances.*
> (pp193–194, 232 and 275)

But nothing else is to be expected (p199). Similarly, many later writers have conjectured that *if* consumer saturation were a fact, or *if* we would value the leisure dividend of efficiency increases more, problematic over-consumption and high natural-resource rebound would be mitigated (Schor, 1992 and 1999; Grubb, 1990a; Sanne, 2000, pp489–490 and 494–495).

Although Say once for some reason writes cautiously that 'the productive agency thus released *may* be directed [*peuvent être employés*] to the increase of production' (p295, emphasis added), only McCulloch took this possibility of non-consumption seriously:

> *If the labourer's command over the necessaries and comforts of life were suddenly raised to ten times its present amount, his consumption as well as his savings would doubtless be very greatly increased; but it is not at all likely that he would continue to exert his full powers. In such a state of society workmen would not be engaged twelve or fourteen hours a day in hard labour, nor would children be immured from their tenderest*

years in a cotton mill. The labourer would then be able, without endangering his means of subsistence, to devote a greater portion of his time to amusement, and to the cultivation of his mind. (pp167–168)

Our epigraph shows the mainstream view that indolence is seldom chosen. To be sure, Mill attributes this 'less leisure' only partly to unlimited desires; rising population and diminishing agricultural returns also figure (p12). And indeed if Malthus's own principle of population is taken seriously, and 'multiplication … may be regarded as infinite', demand for more efficiently produced food and clothing is likely to dominate over the 'power to consume … in idleness' what has already been produced (Mill, pp34 and 154). Smith's view also ran contrary to Malthus's: while the stomach is limited, our further willingness to purchase is not (Smith, I.xi.c.7), and in the end McCulloch himself seconded this without reservation (pp167–178; see also Petty, p307) The doctrine thus stood that 'the limit of wealth is never a deficiency of consumers, but of producers and productive power' (Mill, p68).

For Rae, likewise, 'All instruments exist solely to supply wants' (p166). As proof he offers a psychological theory *why* indolence loses out to accumulation: 'The increased facility of production has … in a great measure also been absorbed by vanity' (p289). While he takes leisure and indolence seriously, and regards labour a cost (pp98, 118, 141 and 209), display consumption wins out (p271); indeed his chapter 'Of luxury' recounts in detail the human tendency towards display, competitive or prestige consumption (Rae, pp265–292), presaging Veblen's famous 'conspicuous consumption' (1899, pp32, 110 and 241; Sismondi, vol 2, p318). This relative consumption is by definition limitless (Alcott, 2004).

Unlike Veblen, Rae quotes extensively from other authors like Pliny, Smith, Heinrich Friedrich von Storch and Say's similar but less systematic analysis in his chapter 'Of individual consumption – Its motives and effects' (Rae, pp401–411). In a nascent appeal for sustainability Rae praises care for 'futurity', 'frugality' and saving in the interests of the 'social affections' (pp60, 265 and 275), strongly seconded by Jevons in his worry for posterity over coal's depletion (pp3–6, 373, 412 and 454–455). But these succumb in great degree to vanity:

At length, in some quarter or another, an improvement began to be perceived. What do we find to have been the most prominent accompaniment of this change? Is it a diminished expenditure – and increased parsimony – a frugality before unknown? I believe not.
(Rae, p23)

Mill even built this power of consumption over investment and indolence into his very definition of political economy, which 'makes an entire abstraction of every other human passion and motive; except those which may be regarded as perpetually antagonizing principles to the desire of wealth, namely aversion to labour, and desire of the present enjoyment of costly indulgences' (quoted by Bladen in Mill, pxxix). Our fifth (no-)rebound category stands as an extreme: at absolute consumer saturation every efficiency increase would bestow upon us free time and upon posterity relatively more resources.

BACKFIRE

Malthus was the economist most worried about market glut or an insufficient 'extension of the market' (pp285 and 288).[76] But he too in the end attested high rebound and even, with regard to labour inputs, *direct backfire* – for instance in the case of cotton goods where 'notwithstanding the saving of labour, more hands, instead of fewer, are required in the manufacture' (p281). He accordingly defended himself against being 'classed with M. Sismondi as an enemy to machinery' (p282 note). Between the first and posthumous second editions of his *Principles*, in 1820 and 1836, many writers had banned thoughts of consumer saturation, if they occurred at all, to the realm of theory. McCulloch recaps the story thus:

> *Accumulation [of capital] and division [of labour] act and react on each other. The quantity of raw materials which the same number of people can work up increases in a great proportion, as labour comes to be more and more subdivided; and according as the operations of each workman are reduced to a greater degree of identity and simplicity, he has … a greater chance of discovering machines and processes for facilitating and abridging his labour. The quantity of industry [labour], therefore, not only increases in every country with the increase of the stock or capital which sets it in motion; but, in consequence of this increase, the division of labour becomes extended, new and more powerful implements and machines are invented, and the same quantity of labour is thus made to produce an infinitely greater quantity of commodities.* (p96; see also Jones, pp237–244)

Three points of note in this passage are as follows: first, McCulloch seems to be considering material rather than labour inputs. Next, circulating as well as fixed 'capital' is endogenized (see also pp94–95; Mill, p63). And third, if material output ('commodities') really grows as much as he says, then backfire is very likely. Babbage likewise discusses efficiency in material/energy as well as time terms, and regards the growing economy as too obvious to mention (see, for example, pp100, 112, 222 and 273; Mill, p106). Rae concurs with McCulloch in almost the same words (Rae, pp67–68).

If McCulloch were to visit us today, would he regard his term 'infinite' as an exaggeration? He would in any case see the understatement in his view that the 'admirable machinery invented by Hargreaves, Arkwright and others [enables] us to spin a hundred or a thousand times as great a quantity of yarn as could be spun by means of a common spindle' (p99). As Rae imagined, were 'some one of the men of olden time, waked from the slumber of the tomb and raised up to us', to witness even a ten-fold yield, 'he might well demand how the power had been acquired that had wrought so great a change' (p14).

Let us take McCulloch literally: without the efficiency granted us by the machines, we would make *much less* yarn. In Jevons's version 'economy renders the employment of coal more profitable, and thus the present demand for coal is increased. ... [I]t cannot be supposed that we shall do without coal more than a fraction of what we do with it' (pp8, 9, 141 and 190). This thought is radical. Today's environmental efficiency strategy claims that an input's more efficient use lowers its rate of consumption. The inverse/corollary of this is that were processes to become *less* efficient, we would consume the input at a *higher* rate. Or had technological efficiency increase remained unchanged – stopped, say, around 1781 with 'the introduction of Watt's engine, the pit-coal iron furnace, and the cotton factory' (Jevons, p270) – we would, according to the strategy's assumptions, today consume a hundred or a thousand times as much – or infinitely more – labour or cotton or fuel than we do today after over two centuries of efficiency increase. To maintain that rebound is less than 100 per cent one must defend this conclusion.

Jevons asks 'could we desire that Savery, Newcomen, Darby, Brindley and Watt' had not increased our industrial efficiency (p457)? Say envisions the case of frozen technology in imagining that a given road exists still as just a path with much less transport efficiency. He says that we can't measure the 'gain' to consumers of the road because with no road 'the transport would never take place at all' (p443 note). Malthus similarly wrote, 'If the roads and canals of England were suddenly broken up and destroyed ... there would be immediately a most alarming diminution both of value and wealth' (p243). As seen above, Jevons's comparable example was that Savery's steam engine '*consumed no coal, because its rate of consumption was too high.* ... It was so uneconomical that, in spite of the cheapness of coals, it could not come into common use' (pp143 and 118; Rae, pp247–248). Marx would later conclude that without machines, for example, '£2000 capital would, in the old state of things, have employed 1200 instead of 400 men' (p393). More drily, Mill takes division of labour as the proxy for improvement in efficiency and notes, 'Without some separation of employments, very few things would be produced at all' (p118).

Say played further with this mental exercise. In connection with his example of printed pages as a case of direct backfire he writes of efficiency-induced price falls that:

> *sooner or later ... cheapness will* run away with *the consumption and demand [and] in all the instances I have been able to meet with, the*

> *increase of demand has invariably* outrun *the increasing powers of an improved production.* (pp87 and 302, emphasis added)

That is, imagine the 'relative intensity of supply and demand', which determines price (Say, p290), as showing flat demand curves and steep supply curves. Now, he said:

> *... suppose ... the charges of production are at length reduced to nothing. ... Every object of human want would stand in the same predicament as the air or the water, which are consumed without the necessity of being either produced or purchased. In like manner as every one is rich enough to provide himself with air, so would he be to provide himself with every other imaginable product.* (pp303–304)

Would total, overall, absolute consumption of resources be lower, or higher, in this state of infinite efficiency, where both commodities and their inputs are free and limitless?

Smith casts some doubt on this, writing that if a 'capital ... was produced spontaneously, it would be of no value in exchange, and could add nothing to the wealth of society' (II.v.5). But Say also takes the exercise in the opposite direction:

> *By the rule of contraries, as a real advance of price must always proceed from a deficiency in the product raised by equal productive means, it is attended by a diminution in the general stock of wealth.* (p302; see also Smith, I.xi.o.6)

That is, is greater wealth even conceivable under conditions of *decreasing* efficiency? If we take time, material, energy and space inputs and assume all historically known efficiencies away, we most likely arrive at the population and per capita production of hunter-gatherer societies living sustainably.

Sarcasm also distinguished an anonymous 1826 article on the 'machinery question' of technological unemployment:

> *If the use of machinery is calculated to diminish the fund out of which labourers are supported, then by giving up the use of the plough and the harrow and returning to the pastoral state, or by scratching the earth with our nails, the produce of the soil would be adequate to the maintenance of a much greater number of labourers. There are many labourers now in England, and the gradations of ingenuity and skill in machinery are numerous; but as the number of labourers and the funds for their support would be gradually increased in proportion as we fell back upon the less perfect machinery, so, at last, when we deprived ourselves entirely of its assistance, the produce and hence the population*

> *of England would be increased beyond what has ever been exhibited in*
> *any country upon the surface of the globe.* (Anon, 1826, p102)[77]

The writer is criticizing Mr Wakefield and Dr Chalmers, but also chides Ricardo
for his change of heart on this question – of which more in the final section.

Say twice frames his description of consumption growth in terms of inputs.
Demand 'outruns' efficiency in a:

> *production, operating upon the same productive means; so that every*
> *enlargement of the power of the productive agency has created a*
> *demand for more of that agency, in the preparation of the product*
> *cheapened by the improvement. ... When the demand for any product*
> *whatever is very lively, the productive agency, through whose means*
> *alone it is obtainable, is likewise in brisk demand, which necessarily*
> *raises its ratio of value: this is true generally, of every kind of productive*
> *agency.* (pp302 and 324)

If the phrase 'ratio of value' refers to amounts of the input before, and after, the
improvement, perhaps times their price per unit, Say is presaging Jevons's
position exactly. Similarly, depending upon one's interpretation of Smith's term
'fund', he too could be attesting rebound greater than unity when he claims that
'Every saving ... must increase the fund which puts industry into motion and
consequently the annual produce of land and labour' (II.ii.25).

As shown earlier, Rae frequently frames his analysis in terms of materials
rather than labour, but he seems usually to denote only the materials *embodied in*
tools, machinery and instruments, as when he speaks of 'the efficiency of ...
materials when formed into instruments' (p112). However, since fields and foods
are also 'instruments', we can infer that efficiency in some cases implies increased
inputs of things other than knowledge (pp112–113): 'Every society possesses a
certain amount of materials capable of being converted into instruments' (pp99
and 187). For Rae, greater efficiency of an instrument means it yields 'quickened'
returns (p164) and, in general:

> *the effect of improvement, to carry instruments into orders of quicker*
> *return ... a greater range of materials is brought within the reach of*
> *[the accumulative] principle, and it consequently forms an additional*
> *amount of instruments. ... All [improvements], therefore, place a*
> *greater range of materials within compass of the accumulative*
> *principle, and occasion the construction of a larger amount of*
> *instruments.* (pp261, 131 and 365)[78]

Furthermore, a 'multiplication of instruments is of no avail, unless something
additional be given on which they may operate', and our 'instruments ... draw

forth stores' of materials; 'improvement in their construction ... put additional stores within reach of the nation' (Rae, pp29, 19 and 68). In addition, 'The various agricultural improvements ... with which invention enriched that art in Britain ... occasioned a great amount of material to be wrought up, which before lay dormant' (p261).

Finally, with a rebound example familiar from today's debate, he notes of the macadamization of roads that 'the facility it gives to transport occasions an increase of transport' (p114). Hearn similarly writes of invention that it 'enables the labourer to work materials which ... were previously beyond his reach' (pp181–183). Taken together these observations are arguably a description of backfire: ultimately, efficiency leads to higher rates of *material consumption*. Since each instrument – a field, a steam engine – implies not only embodied but operating materials, we can infer little saving of material inputs from Rae's analysis. He continues by noting that improved instruments increased the amount of land under cultivation and that 'rocks were quarried; forests were thinned; lime was burned; the metal left the mine' (pp261–262). A rise in Q entails rebound for sure, and most likely backfire.

A summary by Mill contains almost all of the concepts introduced till now. Recall that 'circulating capital' covers all the food, fuel and other materials fed into production. Just before considering the 'stationary state' and 'to what goal ... economical progress' should be aimed (p752), he writes:

> *It already appears from these considerations that the conversion of circulating capital into fixed, whether by railways, or manufactories, or ships, or machinery, or canals, or mines, or works of drainage and irrigation, is not likely, in any rich country, to diminish the gross produce or the amount of employment for labour. How much then is the case strengthened, when we consider that these transformations of capital are of the nature of improvements in production, which, instead of ultimately diminishing circulating capital, are the necessary conditions of its increase, since they alone enable a country to possess a constantly augmenting capital without reducing profits to the rate which would cause accumulation to stop. There is hardly any increase of fixed capital which does not enable the country to contain eventually a larger circulating capital than it otherwise could possess and employ within its own limits; for there is hardly any creation of fixed capital which, when it proves successful, does not cheapen the articles on which wages are habitually expended. All capital sunk in the permanent improvement of land lessens the cost of food and materials; almost all improvements in machinery cheapen the labourer's clothing or lodging, or the tools with which these are made; improvements in locomotion, such as railways, cheapen to the consumer all things which are brought from a distance. (pp750–751; see also p344)*

A few pages later our epigraph appears wherein Mill doubts that any labour had been saved by labour-saving devices. This fruit of classical thought fell to Jevons.

THE PRINCIPLE OF POPULATION

Since the classical economics era, population size seems to have declined in importance as a dependent variable; yet the ten-fold increase of population in the last two centuries is surely an *explicandum* of the first order. No classical economist challenged productivity's causal role. Today, by contrast, this is denied, for instance, by Schipper and Grubb, who, although they 'normalize ... observations of absolute quantities to either population or GDP' see none of this 'significant' population growth as 'stimulated by the increases in energy efficiency' (2000, p368). Perhaps the OECD perspective of almost all studies, abetted by shyness in the face of the fact that people do die from lack of sustenance, has prevented the adoption of both agricultural and manufacturing efficiency as an independent variable. Yet if population rise is at least enabled by efficiency increase, then the wholly exogenous treatment of population in energy-consumption models is wrong (for example Schipper et al, 1996, p174; Howarth, 1997, p4; Lantz and Feng, 2006, p235). It also means underestimation of rebound.

Presaging I = PAT, Jevons made the point that the 'quantity of coal consumed is really a quantity of two dimensions, the number of people and the average quantity consumed by each' (p196). Malthus, in both his major works, endogenized 'number of people'; his metaphorical phrase was that 'the necessaries of life, when properly distributed, [create] their own demand [by] raising up a number of demanders' (p113; see also pp114, 130, 181, 223 and 251). He then points out that if increased 'powers of production' were not necessary for increased population, 'the Earth would probably before this period [mid-19th century] have contained, at the very least, ten times as many inhabitants as are supported on its surface at present' (pp251 and 288). In explaining wealth, '[to] suppose a great and continued increase of population is to beg the question. We may as well suppose at once an increase of wealth' (p252). (Ironically, countless modellers of rebound do exactly this, exogenize GDP, 'economic activity' or total output![79]) As shown earlier, classical economics almost fully endogenized growth, attributing the size of the annual produce of land and labour partly to 'improvement' – as Mill's statement quoted above shows. Progress raises sustenance (in spite of diminishing returns in agriculture), increasing the extent of the market, which in turn allows more division of labour and larger, more expensive machinery, in turn enabling larger population (Mill, pp33, 129–131, 190 and 712–714).

Perhaps building on Petty (p255), Smith states simply, 'The number of workmen increases with the increasing quantity of food, or with the growing improvement and cultivation of the lands' (Ixi.c.7). Building on Say (pp71 and 292–295), McCulloch writes that 'there does not seem to be any good reason why man himself should not ... be considered as forming a part of the national capital. Man is as much the produce of labour as any of the machines constructed by his agency' (p115; see also Mill, pp40–41). Malthus talked of the 'cost of producing a poacher' compared to that of a 'common labourer or ... coal-heaver' (p180; see also Jones, p196). Rae abstractly but explicitly named 'invention' as 'the true generator of states and people' (pp31 and 323). Sustenance includes not only food but warmth, housing and general health (Say, pp301 note, 373 and 378; Mill, pp154–159). The quantity of labour (and people) is a function of the quantity and quality ('human capital') of labour.

Starting with Petty's question as to how many men the land would feed, all of the old-timers embraced the principle of population, expressed by Malthus in terms of 'tendencies', sustenance and the effect of prosperity on decisions to marry and have children (1798, pp20–26, 33–34, 41, 52, 70 and 74–75).[80] Jevons, of course, tied it empirically with coal: '[With] cheap supplies of coal, and our skill in its employment ... [w]e are growing rich and numerous' (pp199–200). In terms of the I = PAT production function for (negative) environmental impact, where Impact results from Population, Affluence and Technology, we should write I = f(P, A, T). A = f(T) shows our becoming rich while P = f(T) shows our becoming numerous. That population is not *sui generis* is also shown and recognized by recent investigators (for example Giampietro, 1994, pp680–681; Hannon, 1998, p215).[81] Schmookler was one who consciously treated it both exogenously and endogenously (1966, pp104–106; Rosenberg, 1982, p141). If, moreover, population and the scale of the economy are partially endogenous, the ubiquitous picture in the literature of a 'race' between a 'growth effect' and efficiency is incorrect (Levett, 2004, p1015).[82] The question of backfire is begged when growth and efficiency are assumed to be rivals, but the race metaphor again shows the *paradox*: do efficiency increases compensate for growth or cause it?

Another population-related problem with most rebound analyses is the concept of the energy intensity of a given good, service or expenditure whereby 'energy costs are typically a small component of the total cost of owning and operating energy-using equipment' (Howarth, 1997, p2). '[T]otal energy costs are generally a few per cent of GDP' and the size of any rebound or 're-spending effect [where] purchasing power is released for other energy-containing services' is proportional to this percentage (Grubb, 1990b, p784; Greening et al, 2000, p391).[83] In analysing indirect rebound, for instance, one compares the energy intensity of the old and the new expenditure to help measure the change in energy consumption. As in Malthus's defence of the concept of natural price, this energy share and the other intensities, for example of labour or capital, add up to 100 per cent (pp66–67).

However, as shown above in discussing Say's 'immaterial objects', buying labour also implies expenditures by the labourers on material and energy, in the older terms of 'reproducing' themselves. Kaufmann's rendering of this 'feedback' effect for capital as well as labour is that when these are substituted for energy, these also have energy costs, which 'offsets some fraction of the direct energy savings and reduces the amount of energy saved by price-induced microeconomic substitution' (1992, p49). Mill's detailed analysis of a loaf of bread, for instance, names bakers, ploughmen, plough-makers, carpenters, bricklayers, hedgers, ditchers, miners and smelters who share the price (costs) of the loaf (p31). Labour and capital, the more so when seen in the classical sense as previous embodied labour, entail energy consumption and are not somehow energy-neutral (Costanza, 1980). Mill also incidentally rejected the implication of perfect substitutability in these analyses:

> When two conditions are equally necessary for producing an effect at all, it is unmeaning to say that so much of it is produced by one and so much by the other; it is like attempting to decide which half of a pair of scissors has most to do in the act of cutting; or which of the factors, five and six, contributes most to the production of thirty. (pp28–29)

In any event, the notion that 'non-energy' costs have no effect on energy consumption must be rejected: once the creation and support of population is included, attending a concert is not the environmentally friendly act it is alleged to be. The idea of decreasing marginal energy intensity as income rises – also due to the societal income effect – must be doubted.

Global population, along with technologically achieved levels of affluence, entailing as they do human usurpation of the living space of plant and other animal species, engenders interest in possible rebounds in the use of a further productive input, namely space, or land regarded merely as m^2 ($\lambda m^2 \uparrow \rightarrow m^2 \uparrow$). Not only agricultural efficiencies, but also transport and architectural ones, can be expressed in terms of amount of land use, raising the question of whether, for instance, more efficient farming reduces the pressure on forests (Jevons, p200; Pascual, 2002, p497). Whenever classical literature raises this question, the answer is that following agricultural improvement we do *not* take land out of cultivation.[84]

THE EMPLOYMENT PARADOX

Because they directly raise population, labour and energy efficiency increases thus indirectly raise the number of work hours or employment; but, given the limited length of the work day, is this true when we hold population constant? Labour rebound would be smaller, but as Mill said most likely work hours don't decrease. Recall that, before Jevons, economists conceptualized all sorts of efficiency changes – not just technological ones – but asked explicitly only after the fate of labour inputs, not of material inputs. Their specific debate concerned whether machines caused long-term unemployment. Jevons of course saw that with 'every ... improvement of the engine ... hand labour is further replaced by mechanical labour' and that in *agriculture* 'labour saved is rendered superfluous' (pp152–153 and 243); also institutional efficiency, through free trade, 'raises the economy of labour to its highest pitch' (p413). But he asserted that it was obvious that demand for labour thereby grew:

> *As a rule, new modes of economy will lead to an increase in consumption according to a principle recognized in many parallel instances. The economy of labour effected by the introduction of new machinery throws labourers out of employment for the moment. But such is the increased demand for the cheapened products, that eventually the sphere of employment is greatly widened.* (p140)

He offers empirical proof with the examples of seamstresses, coal miners and iron workers (pp130–131, 140, 153, 213–218 and 277–278) as his predecessors had with the examples of printing and cottons. As we shall see, this result was not at all obvious for Marx (pp354–392), writing at the same time as Jevons, as it had not been for Ricardo and Sismondi.

The issue is the same as that concerning primary energy: Does an input-saving production system permanently lower, or raise, consumption of that input? We could even call this 'Say's Paradox', for, after demonstrating that cheapened products create additional employment, he writes:

> *Paradoxical as it may appear, it is nevertheless true that the labouring class is of all others the most interested in promoting the economy of human labour; for that is the class which benefits the most by the general cheapness, and suffers most from the general dearness of commodities.* (p89 note)

The result that the 19 out of 20 'unfortunate' men laid off at a flour mill would find other work was for him admittedly '*survenue*' (1820, p63).[85] But he claimed that in printing, even if machines had thrown 199 out of 200 copyists out of work, probably 20,000 people were working in the printing trade (p88).

While many energy efficiency increases cause labour efficiency increases as a side-effect – if only in the mining and distribution of the energy per unit of product – labour-saving changes like new machines, household gadgets or the factory system usually *lower* energy efficiency per unit, if only due to the substitution effect. Such feedbacks between βM and αL could be investigated in complete models of either labour or energy consumption (Rae, p20; Marx, pp386–387; Binswanger, 2001, pp127–128). Again using the example of the ceramic stove replacing the open hearth: heating requires less time cutting and stacking wood as well as less wood (see also Jones, pp249–250; Mill, pp106–107; Martinez-Alier, 1987, p3). Hearn's generalized insight was *both* that 'labour and ... time are free to be applied to other industrial purposes' and that 'the introduction ... of natural forces in lieu of or in addition to human powers sets free a quantity of commodities' (pp183–185 and 271). But the Jevons Paradox concerns only $M = f(\beta M)$, not $M = f(\alpha L)$ as well.

By arguments from price falls, profitability and the income effect, a near-consensus reigned concerning output growth and labour-input growth – epitomized by Mill's quip in our epigraph. Some years before the outbreak of the controversy over machines vs. men, Smith claimed that:

> *the accumulation of stock must ... be previous to the division of labour. ... As the division of labour advances ... in order to give constant employment to an equal number of workmen, an equal stock of provisions, and a greater stock of materials and tools than what would have been necessary in a ruder state of things must be accumulated beforehand. But the number of workmen in every branch of business generally increases with the division of labour in that branch. ... The increase in the quantity of useful labour actually employed within any society must depend altogether upon the increase of the capital which employs it.* (II.intro.3, IV.ix.36)

Remembering that 'capital' is both fixed and circulating (in this case wages in the form of food and provisions during the period of production), and that fixed capital always entails heightened efficiency (Jevons, pp150 and 155), Smith's view is that technological efficiency ('tools') and organizational efficiency ('division of labour') are the conditions for growth in the number of jobs. There is no hint that machines throw people out of work.

However, the intuition that makes the economy of labour just as paradoxical as the economy of fuel, and the fact that visibly and locally machines *do* replace workers, had by the early 1820s spawned the theoretical positions of Say, Robert Owen, Ricardo, Sismondi and Malthus. Say first discussed the displacement of workers in his first edition in 1803 (Chapter IX), making important changes but keeping his conclusions in later editions as well as in the fourth of his *Letters to Malthus* (1820). Lauderdale also explicitly discussed machines that 'supplant labour', first agreeing with Smith that lower labour costs in textile manufacture had lowered prices and that machines generally increase wealth, but at the same time attesting a net loss for the supplanted 'unlettered manufacturers themselves' and seeing good reason for the 'riots that have taken place on the introduction of various pieces of machinery' (pp168–171, 184, 189–192 and 206).

Reminiscent of much microeconomic work on rebound today, most participants traced the fate of the money amounts of capital or revenue saved by efficiency increase. Employment was gained by making and maintaining the machines, but lost when production processes needed fewer hands; it was gained when employers spent their higher profits on luxuries or servants, but lost if demand for other products failed. The monetary examples are found in Ricardo (pp16 and 388–391), Sismondi (vol 2, pp324–326), Say (1820, pp60–61 and 65–67), Malthus (pp192–194 and 282–283), McCulloch (pp179–182) and Marx (pp392–393). The parameters to observe are: 1) percentage labour efficiency increase compared to percentage price fall (usually seen as equal); 2) total fixed capital; 3) total circulating capital shifted between workers in different branches and between workers and capitalists; 4) the income effect of demand for other products; 5) labour demanded for making and tending the machinery; 6) duration of the machine; 7) demand for 'unproductive labour' or 'menial servants', whom these writers do not (usually) count as 'labourers'; 8) foreign demand; and 9) the short-run displacement of labour.

Most of these appear in Ricardo's contradictory discussion. In the third edition of 1821, without explicitly answering Say, he acknowledges a change of mind. Earlier he believed that an increase of 'net income' (rents and profits) always entailed an increase of 'gross income' (including wages and implicitly jobs), arguing in Parliament against Owen's opposite view (Sraffa, 1951, plviii). In his new chapter 'On machinery' he is thinking out loud: because the employer has less 'circulating capital … his means of employing labour would be reduced' (p389); but with increased profits after the introduction of the machine, the 'power of purchasing commodities [of the 'net produce'] may be greatly increased' (pp389–390). In asserting that 'there will necessarily be a diminution in demand for labour [and] population will become redundant', his system boundary remains at the single factory or sector, in other words he forgets indirect rebound (p390); yet due to the necessary 'reduction in the price of commodities consequent on the introduction of machinery … there would not necessarily be any redundancy of people' (p390; see also p392).

He then seems to forget price reductions, doubting the demand, for instance, for a greatly increased supply of cloth (p391). In the simple example of replacing men with horses, he sees a case of 'gross revenue' falling while 'net revenue' rises (p394); yet even here, the income of the farm employer could be so great, or 'the produce of the land [so] increased', that all of the unemployed find jobs 'in manufactures, or as a menial servant' (pp394–395). On the one hand he states:

> All I wish to prove is that the discovery and use of machinery may be attended with a diminution of gross produce … injurious to the labouring class, as some of their number will be thrown out of employment. … [A]n increase of the net produce of a country is compatible with a diminution of the gross produce. … By investing part of a capital in improved machinery, there will be a diminution in the progressive demand for labour. (pp390, 392 and 397)

On the other he believes that 'the employment of machinery should never be safely discouraged in a State [and] that machinery should ... be encouraged' – both because its introduction is slow and because otherwise even jobs in the machinery industry would move overseas (pp395–396). In the terms of today's debate, Ricardo is arguing that rebound is never greater than 100 per cent and tends to be quite a bit less.

Say directly attacks the issue both in his *Treatise* (pp86–90) and in the fourth of the *Letters to Malthus* (1820). In the latter he explicitly bases his case first on large price falls and high price elasticity of demand (1820, pp56–57), second on latent demand for other commodities that is satisfied by the income effect (which he unjustly accuses Sismondi of neglecting) (1820, pp60–62), third on the fact that the machines can simply *do more work* than men (1820, pp58–59), and fourth on the fact that, after all, the factory produces the same amount of product available for consumption, and the laid-off workers, with this sustenance, will do something else (1820, pp61–63). Mill echoed this last point in making the softer claim that 'if there are human beings capable of work, and food to feed them, they *may* always be employed in producing something' (p66, emphasis added). It seems also to be the case today that natural resources not used for one purpose get used for another.

Say goes on to convincingly show that Sismondi's monetary example contains some unrealistic assumptions, but himself makes two numerical errors (Say, pp60–61). He then appeals both empirically to the high and increasing employment all around him (p63) and to a historical overview: his 'model' predicts – accurately – that:

> *if the arts still improve ... they will produce more at less expense [and] fresh millions of men in the course of a few ages will produce objects which would excite in our minds, could we see them, a surprise equal to that which the great Archimedes and Pliny would experience could they revisit us.* (p64)[86]

Two ambiguities mar the comparison of labour with material/energy inputs as well as the classical debate over the former. First, saving material is unmitigatedly good whereas saving labour, because people as opposed to materials must eat, is not. Holding population constant and raising work efficiency, the same or greater employment than otherwise (rebound 100 per cent or backfire) guarantees livelihoods. Somewhat contrary to the view that labour is painful and irksome, rebound greater than unity is therefore good. On the contrary, while resource consumption is obviously good for affluence, its 'over-consumption' and hence backfire is bad due to scarcity and pollution problems.

Secondly, precisely the book-keeping offered by the debate's participants shows that the social or livelihood or full-employment problem is soluble: the amount of output does not decrease! Or, as Ricardo concedes from the point of view of income rather than production, if employers lay off five of ten men, they nevertheless retain the purchasing power to employ all ten (1820–22, p355). If the fully realized production possibilities of the society supported everybody before, it can thus support them after all the great and small productivity increases taking place daily. Therefore even those who held that efficiency savings were in fact realized placed blame on the 'factory' or 'capitalist' set of institutions which included neither shorter work hours nor guaranteed employment. Many such as Owen (Sraffa, 1951, pplvii–lx; Berg, 1980; Greenberg, 1990, pp710–712) and Sismondi (vol 2, pp312–313 and 317) thus mixed ethical or socialist arguments with economic ones. Even Marx, who on the one hand maintained that not only in the short run 'in the hands of capital' labour-saving productiveness increase meant 'lengthening the working day', wrote that:

> ... *workpeople [should] distinguish between machinery and its employment by capital, and to direct their attacks not against the material instruments of production, but against the mode in which they are used.* (pp351, 356 and 374)

His doctrine, though, is that machinery and men are in competition; although new capital can employ many of the newly unemployed and although indeed as much or more 'of the necessaries of life' are still produced, a sufficient rise of demand is uncertain (pp374 and 384–386).

Thus, if the remaining work and/or the same or increased output is distributed equally, the problem of computing the total-employment effects of labour efficiency would lose its social aspect. Again, all agreed with Say's point that even if a wind-driven flour mill does the work of eighteen persons, these 'eighteen extra [redundant] persons are [theoretically] just as well provided with subsistence' (p90; see also Rae, p259). The parallel to energy inputs is that after a machine 'does the work' of one out of two tons of coal, both the coal and the means to employ it remain. And Say, Malthus, McCulloch and Mill, although convinced that even more labour ensued (backfire), recognized that some measures to lessen the hardship of displaced workers are justified. Mill even imagines a 'benevolent government' assuring a just distribution of work, in other words income (p67). A consensus was within grasp that whatever the final level of employment, one must regard full employment as a social, not an economic, problem, as expounded by Edward Bellamy in his *Looking Backward* (1887).

The result is that if produce stays at least the same, 100 per cent rebound in terms of work hours is *possible* at no additional cost. As Malthus claimed, the 'net produce' could always employ 'unproductive labourers' such as 'menial servants, soldiers and sailors' (p191). But the opposite is also possible. In a difficult passage which earned him a reputation as an advocate of labour rebound less than unity, he says that even with increasing 'exchangeable value of the whole produce' stable or sinking employment *could* result, namely when the production of 'luxuries and superior conveniences' rose at the expense of necessaries; but his more fundamental claim is to deny *any* proportional connection between either fixed and circulating capital, and thus efficiency, and demand for labour: consistent with his *Essay on the Principle of Population*, this depends only on 'the means of commanding the food, clothing, lodging and firing of the labouring classes of society' (pp190–191).

If production is higher, some combination of raised affluence and raised population results. If, however, we assume that before the efficiency increase every worker was working his maximum number of hours, then without population increase, labour backfire is logically impossible (Malthus, pp62–63). (Analogous energy-rebound limits perhaps exist through scarcity or thermodynamic limits.) Malthus in fact concludes that if the 'introduction of fixed capital' is gradual and 'the funds destined for the maintenance of labour' somehow keep pace, the result is a 'great demand for labour and a great addition to the population [and] there is no occasion therefore to fear that the introduction of fixed capital ... will diminish the effective demand for labour' (p193; see also pp281–289). By 1836 he accordingly defends himself against being 'classed [by McCulloch] with M. Sismondi as an enemy to machinery' (p282 note), also rejecting the doubts of Ricardo and the opinions of 'M. Sismondi and Mr Owen' that labour-saving machines are 'a great misfortune' (p295 note).

McCulloch was indeed just as convinced as Say that the 'extension and improvement of machinery is always advantageous to the labourer' (p165), but not only because more work hours result. His first original point is that if machinery lowers demand for labour by raising labour's productivity, then so would 'improvement of the science, dexterity, skill and industry of the labourer'; therefore 'M. Sismondi could not ... hesitate about condemning such an improvement as a very great evil' (pp165–166). As seen above, McCulloch's macroeconomic assumption of a ten-fold efficiency increase would allow more leisure (pp166–168; Mill, pp105–106). His result entails considerable rebound in material/energy consumption; there is no backfire in labour consumption but rather real savings of labour inputs; and the imagined cornucopia would enable society to politically assure full employment.[87] But he assumes no population growth. If population and/or work hours increase, L-backfire could ensue.

Microeconomically McCulloch argues explicitly with the standard price falls, large price elasticities of demand and indirect rebound (pp176–180). In apparent contradiction to his vision of shorter working hours for all he then relies on both theory and observation to show that the machines of 'Hargreaves, Arkwright and Watt' created employment for 'thousands and thousands of workmen' (p117). This raises our paradox again: according to Dolores Greenberg, the Owenite John Brooks calculated in 1836 that machines in Great Britain and Ireland were doing the work of no fewer than 600,000,000 people (Greenberg, 1990, p711; Jevons, p411). Can we infer from this that therefore 600,000,000 people were out of work – perhaps even in the sense that they had starved or not been born? If the machines were doing the work of only 300,000,000 people, would employment be twice as high?

Some of Jevons's statistics on population and substitution hint at these questions:

> *In round numbers, the population has about quadrupled since the beginning of the 19th century, but the consumption of coal has increased sixteen-fold, and more. The consumption per head of the population has therefore increased four-fold.* (p196)

Pertinent to today's 'renewables' discussion, he computes, for instance, that since an 'ordinary windmill has the power of about 34 men, or at most seven horses ... the great Dowlais Ironworks ... would require no less than 1000 large windmills!' (pp164–165; see also pp203–205) And when he writes that 'it cannot be supposed that we shall do without coal more than a fraction of what we do with it', we may ask both how many are in this 'we' (p9) and how well-off we would be, since 'with coal almost any feat is possible or easy; without it we are thrown back into the laborious poverty of early times' (p2).[88]

Say, Malthus and McCulloch do not show labour backfire with *certainty*. They show us not that more work hours must result, but that fewer work hours must not result. Even Sismondi saw cases when, for instance, workers were not 'rendered superfluous' due to the stocking-machine – but only because of the three exogenous factors 1) changes of taste, 2) increased population and 3) increased wealth (vol 2, pp316–317 and 330–331).[89] But in the normal case, and contrary to Say's claims in ridiculing him (1820, pp61–62), Sismondi does say that the stockings are cheaper and that demand can therefore rise due to the income effect in sectors having nothing to do with the one affected by the efficiency increase; but he treats the total purchasing power as no greater than that spent on the more expensive spats previously or even as less: 'new demand will never have the same proportion as that thereby lost by the laid-off workers' (vol 2, pp317 and 322–324; see also McCulloch, pp186–187). A further lack of certainty marks Say's empirical claims: perhaps backfire in cottons and printing is proven, given a demand function, but these are mere sectoral studies with no necessary economy-wide implications (p57).

One of Sismondi's arguments for low rebound is that while a machine may lower *labour* costs by 99 per cent, since the price of stockings consists of more than just labour costs, the price cannot fall in the ratio of the laid-off workers (vol 2, pp323–324). Again, many argue today that since energy costs are only a fraction of GDP, the efficiency elasticity of price is low (Howarth, 1997, pp2–3; Allan et al, 2006, pp18–19). Although this argument loses force if rebound is measured as a percentage not of total economic activity but only of potential engineering savings, its plausibility is a reason why the Jevons Paradox is a paradox. If prices fall 50 per cent there is nevertheless more real purchasing power in the economy, whether the efficiency of a given input rises 51 per cent or 99 per cent; perhaps the concept of the efficiency elasticity of price compares apples and pears.

Mill, finally, confronts the problem we named earlier that the purchasing power drawn to the cheaper, more efficiently produced goods is lacking for the older, previously purchased goods, thus lowering employment in those sectors. On the one hand he attests that:

> *Every addition to capital gives to labour either additional employment or additional remuneration. ... If it finds additional hands to set to work, it increases aggregate produce: if only the same hands, it gives them a larger share of it; and perhaps even in this case, by stimulating them to greater exertion, augments the produce itself.* (p68; see also p87)

But he adds that the standard argument – greater employment through cheaper goods through more efficient production through applying fixed and circulating capital to this sector:

> *does not ... have the weight commonly ascribed to it. ... [I]f this capital was drawn from other employments, if the funds which took the place of the capital sunk in costly machinery were supplied not by any additional saving consequent on the improvements, but by drafts on the general capital of the community, what better were the labouring classes for the mere transfer? In what manner was the loss they sustained by the conversion of circulating capital into fixed capital made up to them by a mere shifting of part of the remainder of the circulating capital from its old employments to a new one?* (p96)

Mill seems here to envision a zero-sum process, which indeed the economy is if measured monetarily with constant money supply. Perhaps his premise is wrong that the capital must be drawn from other, previous employments rather than from the real increased produce or 'returns' per unit of input. This is the answer Say would have given and that Rae gave (p118). Although Mill's subsequent attempt to counter his own argument is unsuccessful he then concludes with Say that employment is not threatened after all but in the end increased (pp119–120, 133–134 and 749–751).

Today no one either hopes or fears that labour efficiency increases do not backfire. It is accepted that for over two centuries such 'improvements' have been accompanied by rising employment and population. A causal connection is even often explicit: more efficiency of all sorts, such as free trade, lower transactions costs, restructuring for synergies in industry and everyday streamlining of work processes, is known to further the economic growth upon which an expanding job market depends. But material/energy inputs are perceived differently, with different goals and hopes. Just as the older debate was fraught with the ambiguity of 'labour' as a cost and 'labour' as a proxy for 'income', today's debate contradictorily lauds efficiency of any sort as a tool for lower environmental impact as well as for growth and affluence. If, however, energy rebound is close to or greater than unity, environmental ends are better served by direct means such as taxation or rationing (Hannon, 1975; Sanne, 2000, pp488 and 491–492; Fawcett, 2004; Simms, 2005).[90]

CONCLUSIONS

Jevons opened his seminal chapter on fuel 'economy' (his term for the efficiency ratio) by quoting Justus von Liebig, who wrote:

> *Cultivation is the economy of force. Science teaches us the simplest means of obtaining the greatest effect [output] with the smallest expenditure of power [input], and with a given means to produce a maximum of force. The unprofitable exertion of power, the waste of force in agriculture, in other branches of industry, in science or in social economy, is characteristic of the savage state, or of the want of true civilization.* (von Leibig, 1851, p462)[91]

Then, as now, force and therefore affluence and civilization lies in fossil fuel. But pollution and pending scarcity reveal the dark side of the prosperity that we so welcome. Roughly in the order of the sections presented above, some conclusions can be drawn on whether more efficiency, *ceteris paribus*, achieves not only affluence and greater population but environmental relief.

Efficiency is an attribute of humans and other natural agents, as well as capital and organization, but is always an output/input ratio. Seeing efficiency increase as larger output, as the classical economists usually did, biases us to find high rebound plausible; seeing it as smaller input biases us toward low rebound and real savings. The term 'rebound' itself is a metaphor describing a bouncing ball, but a bounce all the way into the backfire zone unfairly implies *perpetuum mobile* or more. An analysis of energy consumption is possible without computation of engineering savings derived when one holds consumption constant, and thus without the concepts of rebound and backfire.

In regression analysis, to explain increasing (rates of) energy consumption an independent variable 'technological efficiency' could be taken. But how is this measured for all sectors, all economies, over time and integrating new products? An adequate aggregate metric, whether in monetary, utility or physical terms, is hard to come by, but its absence makes empirical research difficult. The environmentally most relevant path of measuring output physically must seek a metric free of the anthropocentricity implied in terms such as waste, usefulness, quality, service and value, for these conflate environmental with affluence criteria.[92] Rather unscientifically, though, we all assume that technological efficiency continually increases. The classical economists also attested this and correlated it not only with growing production of wealth but sometimes with growing labour and material input quantities. Jevons, for instance, offers the empirical evidence for backfire that alongside great rises in coal consumption, population and affluence there were increases in the economy of fuel, for example in pig iron production by a factor of about seven in 35 years (pp145, 196, 261–271 and 387–388; see also Martinez-Alier, 1987, pp86).

Fruitful empirical research must be at a large enough scale to capture not only indirect rebound in all sectors but also an economy's consumption of imported embodied energy (Jevons, p317). This need to ultimately cover all sectors and economies has been acknowledged.[93] As McCulloch said, we must investigate efficiency effects 'in a country surrounded by Bishop Berkeley's wall of brass' (p185), a good description of the whole globe. The more so since environmental problems are global, our studies should be both global and measure total rather than merely direct rebound.

But in the absence of hard empirical results we must resort to theory, and indeed both sides in today's debate over the environmental effects of efficiency claim 'counterfactually' what energy consumption *would have been otherwise*, in other words without efficiency increases (Khazzoom, 1980, pp22 and 31; Howarth, 1997, p3; Brookes, 2000, p356; Moezzi, 2000, pp525–526; Schipper and Grubb, 2000, p370). Which model, then, better predicts this correlation? That of Jevons can perhaps be quantified as containing a technological rebound factor of slightly over 100 per cent, or an efficiency coefficient in a production function of, say, 1.01. Holding all other variables constant, this model predicts the increase in energy consumption better than models assuming rebound less than unity; these yield a large gap between predicted and real consumption, a gap usually filled by exogenous GDP. Such models must, moreover, show what the causes of increased consumption then in fact are, if not efficiency increases.[94] And these causes must be strong enough to *overcome* the alleged consumption-reducing effect of greater efficiency.[95]

Efficiencies of all provenances have continually expanded the world economy's *production possibilities frontier* and thereby its consumption frontier. Grasping this physically – including the physical inputs into this consumption – can avoid some of the difficulties arising in microeconomic monetary analysis in terms of income effects and societal purchasing power. Yet while this immediately renders large rebound plausible, to directly infer backfire would beg our entire question; the Jevons *Paradox* must be taken seriously. In any case, no answer can do without assumptions or empirical evidence concerning the (non-)saturation of material desires and the effect of greater production on population size.

The policy situation is remarkable. The likelihood that theoretical and real input savings are identical is zero; some rebound is uncontested, and the lowest macroeconomic total-rebound estimates lie in the range of 25–40 per cent. It is therefore truly astonishing that, with a handful of exceptions,[96] government agencies and policy assessment companies do *not* correct for it,[97] but rather in a purely 'engineering' approach set real savings equal to technologically possible savings. However, a rebound coefficient of 0.5, which is at the present state of knowledge justifiable, would significantly alter estimates both of efficiency's effectiveness and of its cost-effectiveness.

Remarkably, Smith's 'human stomach' passage – written about 230 years ago – contains practically all the concepts needed to approach our question:

> *But when by the improvement and cultivation of land the labour of one family can provide food for two, the labour of half the society becomes sufficient to provide food for the whole. The other half, therefore, or at least the greater part of them, can be employed in providing other things, or in satisfying the other wants and fancies of mankind. Clothing and lodging, household furniture, and what is called equipage are the principal objects of the greater part of those wants and fancies. The rich man consumes no more food than his poor neighbour. In quality it may be very different, and to select and prepare it may require more labour and art; but in quantity it is very nearly the same. But compare the spacious palace and great wardrobe of the one with the hovel and the few rags of the other, and you will be sensible that the difference between their clothing, lodging and household furniture is almost as great in quantity as it is in quality. The desire for food is limited in every man by the narrow capacity of the human stomach; but the desire for the conveniences and ornaments of building, dress, equipage and household furniture seems to have no limit of certain boundary. Those, therefore, who have the command of more food than they themselves can consume, are always willing to exchange the surplus, or, what is the same thing, the price of it, for gratifications of this other kind. What is over and above satisfying the limited desire is given for the amusement of those desires which cannot be satisfied, but seem to be altogether endless. The poor, in order to obtain food, exert themselves to gratify those fancies of the rich, and to obtain it more certainly, they vie with one another in the cheapness and perfection of their work. The number of workmen increases with the increasing quantity of food, or with the growing improvement and cultivation of the lands: and as the nature of their business admits of the utmost subdivisions of labour, the quantity of materials which they can work up increases in a much greater proportion than their numbers. Hence arises a demand for every sort of material which human invention can employ, either usefully or ornamentally, in building, dress, equipage or household furniture; for the fossils and minerals contained in the bowels of the earth; for the precious metals and the precious stones.* (I.xi.c.7)[98]

Here we find efficiency as 'improvement' and 'division of labour', greater output and expanded production frontier as food surplus, greater population seen endogenously, the irrelevance of the energy proportion of a service, the reduction of quality to quantity, the limitlessness of latent demand, marginal consumers, the empirical fact of consumption going hand in hand with efficiency, and the derived large demand for material inputs including fossil fuel.

Greater technological efficiency enables us to squeeze more useful material out of a given amount of input, or more non-work time out of the 24 daily hours (Sanne, 2000, pp487 and 494). This is Jevons's state of 'happy prosperity' (p276). But if it simultaneously increases demand for natural resource inputs, we face a trade-off between affluence and sustainability. With the evidence at hand today, and given a certain urgency in finding an answer, good judgement is called for. If asked by policymakers today whether we can count on greater energy efficiency to lower energy consumption, how many economists can answer with a whole-hearted 'Yes'?

ACKNOWLEDGEMENTS

Thanks to Len Brookes, Marcel Hänggi, Ashleigh Hildebrand, Reinhard Madlener, Cecilia Roa, Christer Sanne, Irmi Seidl, Steve Sorrell, Steve Stretton, Peng Wang, Özlem Yazlık and the staff of the Zentralbibliothek Zürich.

NOTES

1 Meadows et al, 1972.
2 In our epigraph Mill is stating that labour-'saving' production processes have led to greater demand for labour: with α as an efficiency coefficient, $\alpha L \uparrow \rightarrow L \uparrow$. With this passage from Mill, Karl Marx opened his chapter 'Machinery and modern industry' (1887, p323) and Thorstein Veblen broke for the only time his rule of not quoting or citing anybody (1899, ppx and 111). Jevons's claim, taking E for fuel and β as its efficiency coefficient, is that $\beta E \uparrow \rightarrow E \uparrow$ or $E = f(\beta E)$.
3 The only major challenge known to me is that of Mundella (1878).
4 After granting the physiocrats a germ of truth concerning the priority of land-product surplus, Smith allows himself a joke at their expense (and perhaps that of the present elucidators of the Jevons Paradox): 'as men are fond of paradoxes, and of appearing to understand what surpasses the comprehension of ordinary people, the paradox which it maintains, concerning the unproductive nature of manufacturing labour, has not perhaps contributed a little to increase the number of its admirers' (IV.ix.37–38).
5 The named years of publication are those of first editions, cited here except for Say (4th edition, 1819), Ricardo (3rd edition, 1821), Sismondi (2nd edition, 1827), Malthus (2nd edition, 1836) and Jevons (3rd edition, 1906). *These dates are understood and omitted in all references.* If other writings by these authors are cited, the date is given in the parentheses, for example (Malthus, 1798) or (Say, 1820).
6 Most Environmental Kuznets Curve (EKC) studies suffer the fatal flaw of showing ratios on the vertical axis; for critiques see Jänicke et al, 1989; Opschoor, 1995; De Bruyn and Opschoor, 1997; Alcott, 2006, Section 3.5; Luzzati and Orsini, 2007; Giampietro and Mayumi, this volume.

7 This belongs to our *ceteris paribus* just as did Malthus's two 'postulata' for his principle of population, namely that we need food and that there is passion between the sexes (1798, p19). And it was Malthus who insisted that, following a labour-efficiency increase, we could always choose 'indolence' (p258).

8 See, for example, Jevons, pp85, 91 and 256; Schurr and Netschert, 1960; Cleveland et al, 1984; Schurr, 1985; Smil, 2003, pp6–14, 22–34 and 82–88.

9 For example stemming from education, training, increased effort, Taylorite factory-floor organization, free trade, scientific norms, private property and further cutters of transaction costs.

10 For empirical *sectoral* correlations see Jevons, pp193–194, 232, 275, 154 and 387–388; Greenhalgh, 1990; Rudin, 2000; Dahmus and Gutowski, 2005; Fouquet and Pearson, 2006; Herring, 2006.

11 The causes of efficiency, however, lie perhaps ontologically in capital or organization: the piston, the hot blast and the factory system changed, not coal or iron ore or human beings. Yet classically capital was usually reduced to labour and land, as insisted upon also by Schumpeter (1912, pp20–21, 29, 37 and 210–219); this historical topic is the subject of work in progress. See, for example, Smith, II.ii.25, 33–34; Say, p293; Rae, pp91, 256 and 258; Mill, pp100, 154 and 182.

12 *Sufficient* consumer behaviour, like consumer and production efficiency, also suffers from rebound (Alcott, 2007).

13 Saunders in passing quotes Solow that 'it's hard to break the habit ... "factor-augmenting" does *not* mean "factor saving"' (1992, p131).

14 As shown later, this income effect for consumers, if expressed monetarily, could be balanced by a 'loss effect' for producers.

15 Say spoke for all economists before and since in attesting the disutility of work: 'labour ... implies trouble (*une peine*)' (p85; see also Smith, I.v.4 and I vi.2; Mill, p25). Veblen made fun of our seeming love of 'irksome' labour (1899, ppix, 18–19 and 110).

16 Also Say, pp61–62; Rae, pp1, 15 and 21.

17 Occasionally Smith explicitly inserted 'capital' as input, some given amount yielding 'greater produce', adding K and γK to the production function (IV.ix.6; see also Mill, pp100 and 154).

18 This example reveals further outcomes complicating rebound research: 1) the 'saved' firewood can be used for building and is thus not saved; 2) the time 'saved' cutting and stacking wood can be spent for other earning and consumption.

19 Also Jevons, p177.

20 Also Jevons, pp119, 159 and 389.

21 Rae then offers a full-blown analysis in terms of the varying 'capacities' and speed of returns of tools and machines, a function of their cost of production, their durability and their efficiency (pp87–110), closely resembling that of Malthus (pp71–73). See the analysis of Spengler (1959).

22 Also Jevons, p188; Schumpeter, 1912, pp297–306. Jevons likewise gives many examples of the enlistment of new agents, as opposed to 'subsequent steps in ... improvement' (p119; see also pp113–134 and 147–148).

23 Jevons, pp125–130, 141–144, 152–156, 196–199, 245, 368–378 and 405; Sieferle, 2001, pp115–124.

24 'When we want to double the produce of a field we cannot get it by simply doubling the number of labourers.' (Jevons, p195; see also Smith, I.intro.1 and 5, I.viii.57, II.intro.4, II.iii.32, IV.ix.34; Say, pp70–71 and 303; Mill, pp154 and 413–414).

25 Like McCulloch (pp92–95), Rae took this idea to what he admitted to be an extreme, defining his key concept of 'instruments' to include almost everything having social ontology (resulting from man), including not only tools as conventionally understood but also fields, horses and even food as a means of maintaining human capital (pp86–88 and 115). Although Mill adopted this broad definition for capital, he, like Rae, knew it was too broad for 'general acceptance' (Mill, pp10 and 153).

26 Petty's comparable example had been that 'a mill made by one man in half a year will do as much labour as four men for five years together' (p256).

27 'Life cycle' aspects as well as recycling are thus reducible to our output–input efficiency, as demonstrated by Rae in showing that a more expensive but more durable hat saves labour input for the wearer over time (pp200–201). He also gives examples of thick sturdy walls for buildings and good steel for tools, which *both* increase heating or cutting efficiency and last longer (pp109 and 114).

28 Of course, while McCulloch was asking after the effects on quantity of output (Q), believing 'the power of production ... a thousand or million times increased', the Strathclyde group was asking after the effects on the quantity of consumed *input* once it is used 5 per cent more efficiently.

29 Also Malthus, 1824, p303; McCulloch, p99; Sanne, 2000, p487.

30 Mill added precision to Ricardo's (p80) two types of agricultural improvements, naming some that 'have not the power of increasing the produce', but only diminishing labour (Mill, p180); these *cannot* raise total output of the farm – here the ratio is output/farm – just as some factory-floor efficiencies might increase not the productivity of the factory unit but only that of the labour units.

31 Also Robinson, 1956, p18; Radetzki and Tilton, 1990, p21; Manne and Richels, 1990; Saunders, 2000a, p442; Alcott, 2006, Chapter 6.

32 See Howarth, 1997, p3; Wirl, 1997, p14; Berkhout et al, 2000, p427; Saunders, 2000b; Binswanger, 2001, pp120–121; Sorrell and Dimitropoulos, 2006, p3.

33 Sorrell and Dimitropoulos, 2006, pp3–9.

34 For example Ayres, 1978, pp53–66; Birol and Keppler, 2000, p461; Ayres and van den Bergh, 2005, pp102–103; but see also Weisz et al, 2006, p681.

35 For example Cleveland and Ruth, 1998, p35; van den Bergh, 1999, pp551 and 559; Dahlström and Ekins, 2006, pp509 and 515–518.

36 Also Solow, 1957, pp316–317; Rosenberg, 1982, pp23 and 55; Victor, 1991, pp204–206.

37 And Cantillon's (p2).

38 Also McCulloch, pp61–63; Rae, pp15 and 81–83; Mill, pp25, 27 and 46.

39 Also Ayres, 1978, pp39–66; van den Bergh, 1999; Birol and Keppler, 2000, p461; Schipper and Grubb, 2000, p369.

40 The terms for mass and measure in German are very close (*Masse, Mass*); 'pound' in English is both weight and money, as is *peso* in Spanish (Smith, I.iv.10).

41 Mill distinguishes between the 'absolute waste' of 'unproductive labour' lacking even the utility of 'pleasurable sensation', and the relative waste of 'productive labour' when, for instance, 'a farmer persists in ploughing with three horses and two men ...

when two horses and one man are sufficient' (pp50–51; see also p28; Say, pp42–43, 121 and 404; Alcott, 2004, pp770–776).

42 Mill implies a broader array of formal expressions for efficiency when talking of greater produce 'without an equivalent increase of labour' (p180): the term 'equivalent' implies elasticities, in other words efficiency also increases, for example, in the extreme case where both input and output go down, but the former percentage-wise more than the latter.

43 For example Petty, pp256, 261–264 and 300; Smith, I.xi.o.1, IV.ix.17 and 34–35; Say, pp127, 286 and 432–438; Rae, pp29, 310 and 327; Mill, pp87–88, 133–135, 184–189, 706 and 723; see also McCulloch, pp73–143.

44 See Smith, I.viii.18, 23 and 39, IV.ix.12; Malthus, pp61, 130 and 180; Mill, p33; Jevons, p213; Giampietro, 1994.

45 For example Smith, I.viii.21, IV.ii.9, IV.ix.38; McCulloch, p99; Rae, p7; Mill, p159.

46 For example Smith, II.iii.32, IV.ix.34; Malthus, p252; Rae, pp12–13; Marx, p358; Solow, 1957.

47 For example Smith, I.viii.3, III.i.1; Ricardo, pp273–274; Say, pp71, 86 and 295; Malthus, p296; McCulloch, pp97–102, 166–167 and 411; Jones, pp237–250; Rae, pp15, 99, 216 and 253; Mill, pp88 and 98.

48 Also McCulloch, pp187–188; Mill, pp133–134.

49 For example Smith, I.xi.c.7, II.ii.23; Say, pp240–248; Ricardo, pp274–275; Sismondi, vol 1, pp373–387; Malthus, pp97 and 255; Mill, pp71–72 and 410; also Robinson, 1956, pp18, 24, 65 and 122; Binswanger, 2006.

50 Roughly, 'real', 'inherent' or 'natural' prices were long term and determined by costs of production, while 'market' prices were shorter-term results of supply and demand only; 'nominal' prices were in terms of money (gold and silver). See Mill's 'necessary price, or value' (p471).

51 Also Jevons, pp120, 140, 154, 156 and Chapter V.

52 For example Say, pp300 and 303; Ricardo, pp25 and 52; Malthus, pp281–282; McCulloch, pp117, 176 and 278.

53 In such passages from Smith, Say, Ricardo and Malthus several questions are often discussed simultaneously: 1) why and how wealth increases, 2) how it is distributed between rent, wages and profits, and 3) how supply, demand and price interact in the short term.

54 Also Say, p300; Jevons, pp8 and 140–142; Schumpeter, 1912, pp297–306.

55 Also Mill, pp133–134; Hotelling, 1931, p137.

56 Grubb cryptically adds that 'When energy price or availability constrains demand ... the apparent savings from using more efficient technologies would be largely offset by the macroeconomic response – the tendency to use more energy services because they are made cheaper.' (1990b, p783) That is, he attests very large rebound in run-of-the-mill cases.

57 Say indeed calls 'prix' a measure of 'valeur', and 'valeur' a measure of 'utilité' (p62). But if prices reflect utility, and utility is very different from costs of production, then prices confuse environmental analysis. Utility is not an environmentally relevant concept. If Mill is right, however, that prices in their long-run movement to 'natural price' reflect utility to perhaps 1 per cent and efficiency (or 'difficulty' or cost of

production) to 99 per cent (pp462–464), then this objection falls and prices are a satisfactory proxy for environmental impact.

58 For example Wackernagel and Rees, 1996, pp127–128; Wirl, 1997, p41; Binswanger, 2001, p120.

59 Also Malthus, pp190–192, 296, 319–322 and 339; Jones, pp237–239; Babbage, pp112 and 232–233; Rosenberg, 1982, p106.

60 If the whole mass is X + Y, where X is the newly more efficiently produced good and Y is all else, then $\Delta P \times Q_x$ would equal $\Delta P \times Q_y$.

61 Perhaps Mill's father James led Ricardo to the distinction between the 'net produce' or 'riches', which always increase with efficiency, and the other 'value of that net produce' (P × Q), which 'may not ... increase' (pp16 and 391–392), leaving Mill the work of deciphering.

62 Efficiency and its consequences can be grasped physically. Smith resorts to this method in solving the paradox that 'improvements in ... productive powers' are accompanied not only by price falls but 'in appearance' price rises of many things including labour (I.viii.4; also I.i; Malthus, p215).

63 Saunders shows that backfire is consistent with constant prices when the productivity of energy rises in a production function with capital, labour and energy (1992).

64 The term 'purchasing power' is explicitly found in, for example, Smith, I.v.3, I.xi.m.19–20, II.ii.21; Malthus, pp42, 49, 53 and 80; McCulloch, pp171 and 177; Mill, pp67 and 458.

65 For example Smith, I.vi.6–18; Say, pp15 and 77; Malthus, Book I, Chapters III, IV and V; Mill, p235.

66 Also Mill, pp477–487; Khazzoom, 1980, pp22–24.

67 Ecological economics here differs from Say, who declares these 'spontaneous gifts of nature ... neither procurable by production nor destructible by consumption' to lie outside the realm of political economy (pp63 and 86). In the frequent classical emphasis on exchange, as in environmental economics' emphasis on allocation, one sees that new biophysical facts, and limits, necessitate a redefinition of political economy (Boulding, 1966; Daly, 1992).

68 Say also noted that efficiency is the *result* of a profusion of taxes (p473), a point likewise clear in today's debate, wherein Pearce, for instance, notes that through efficiency some of the effect of eco-taxes is 'taken back' (1987, p14).

69 A friend of mine who wholesaled slide-rules once had to throw away several thousand of them upon the advent of calculators – a process difficult to integrate into this gain/loss calculus and again raising the question of *undesired* output or waste.

70 Brookes concurs with Jevons that really saving such a material lowers affluence (Brookes, 1990 and 2000).

71 See Sanne, 2000, pp488–489; Binswanger, 2001, p122 note.

72 Also Khazzoom, 1980, p32; Grubb 1990a, pp235 and 195; Rosenberg, 1994, pp165 and 166; Schipper and Grubb, 2000, pp368, 383 and 387; Sorrell and Dimitropoulos, 2006, p3.

73 Say at times also eschewed empirical study (p102 note), a view shared less categorically by Ricardo (1820–1822, pp362–363).

74 Also Howarth, 1997, pp4 and 7; Schipper and Grubb, 2000, p384.

75 His claim is also empirical: 'experience amply shows' this (pp284 and 268).

76 Mill also asked who would buy the 48,000 pins now produced every day in Smith's factory, going on to name some conditions for a large market including population and transportation infrastructure (pp129–130).

77 Attributed by Mill to William Ellis (Mill, p736).

78 Also Mill, pp725–726; Price, 1998; Wirl, 1997, pp51–56 and 81–87.

79 For example Manne and Richels, 1990, p51; Schipper and Meyers, 1992, pp58–60; Howarth, 1997, p2; Saunders, 2000a, p442; Schipper and Grubb, 2000, pp368 and 370.

80 Also Cantillon, pp43–44; Smith, I.viii.21–39, I.xi.b.1 and c.7, IV.ix.36; Ricardo, p16; Say, pp189, 322, 371–381 and 450; McCulloch, p278; Rae, pp28–31, 96, 160 and 324; Mill, pp153–159 and 187–190; Jevons, pp222–225 and 420.

81 Also Cipolla, 1962, pp49–53, 94–95 and 105; Martinez-Alier, 1987, pp99–116; Abernethy, 1993; Pimentel et al, 1994; Bartlett, 1994; Clapp, 1994; Johnson, 2000; Giampietro and Mayumi, 2000.

82 Also Besiot and Noorman, 1999, pp375–377; Binswanger, 2001, p120; SwissEnergy, 2004, pp3–4.

83 Rebound should, however, be defined as a percentage of engineering savings, not of GDP.

84 Smith, I.xi.b.2–6, IV.ix.5–6; Say, p295; Malthus, pp139–140; Jones, pp196 and 242; Rae, pp116, 259 and 261; Mill, pp173–185 and 724–729.

85 Curiously, this term is left out of Laski's English translation (p63).

86 He praises the relief from toil offered by machinery (p64).

87 See Bellamy, 1887.

88 That agricultural productivity increases raise population is clear; manufacturing and fuel-using efficiency increases do this less obviously through better housing and clothes, better medicine, better availability through transportation, etc (Jevons, pp200, 205, 233, 243–245 and 369).

89 Just like the very similar independent variables of Schipper and Meyers (1992) and Schipper et al (1996), Sismondi thus begs several questions.

90 Jevons however repeatedly notes that such answers to the coal question are limited by Britain's 'system of free industry' (pp5; see also xlix, 13, 136 and 442–447).

91 Jevons here misquotes von Liebig as 'civilization is the economy of power' (Jevons, pp142 and 163). Jevons had just finished his chapter attributing Great Britain's greatness to coal and technology, whereas von Liebig was in the middle of an essay on agricultural productivity.

92 A given CO_2 molecule, for instance, has no marker on it indicating its human value.

93 For example Saint-Paul, 1995; Cleveland and Ruth, 1998, pp44–45; Giampietro and Mayumi, 2000, pp182 and 185–186 and this volume; Weisz et al, 2006, p694; 4CMR, 2006, pp24 and 52–53; Rhee and Chung, 2006; Polimeni, this volume.

94 See Saunders, 2005.

95 See Howarth, 1997, pp2–4 and 7; Schipper and Grubb, 2000, p384; Solow, 1970, pp33–35 and 38.

96 Rebound coefficients crop up in Defra, 2002, p4; NRC, 2002, sections 4.1 and 5.24–25; 4CMR, 2006, pp5, 12, 21, 35 and 72–75.

97 EEB, 2000, p32; INFRAS, 2003; CEPE, 2003, pp6, 32, 35, 44 and 55; DTI, 2006, pp36–60 and 149; EnergieSchweiz, 2007.

98 Also Ricardo, p293; see Say on cheaper corn and 'dress and household furniture' (p301).

REFERENCES

Where two publication dates are given, the first represents the year of first publication and the second gives the year of the edition cited.

4CMR (Cambridge Centre for Climate Change Mitigation Research) (2006) 'The macro-economic rebound effect and the UK economy', final report to Defra, 18 May, www.defra.gov.uk/environment/climatechange/uk/energy/research

Abernethy, Virginia (1993) 'Why the demographic transition got stuck', *Population and Environment*, vol 15, pp85–87

Alcott, Blake (2004) 'John Rae and Thorstein Veblen', *Journal of Economic Issues*, vol 38, no 3, pp765 786

Alcott, Blake (2005) 'Jevons' Paradox', *Ecological Economics*, vol 54, no 1, pp9–21

Alcott, Blake (2006) 'Assessing energy policy: Should rebound count?', dissertation for the degree of Master in Philosophy in Land Economy, Cambridge University, UK, www.blakealcott.org

Alcott, Blake (2007) 'The sufficiency strategy: Would rich-world frugality lower environmental impact?', *Ecological Economics*, forthcoming

Allan, Grant, Hanley, Nick, McGregor, Peter G., Swales, J. Kim and Turner, Karen (2006) 'The macroeconomic rebound effect and the UK economy. Final report to Defra', www.defra.gov.uk/environment/climatechange/uk/energy/research

Anonymous (1826) 'Effect of the employment of machinery &c. upon the happiness of the working classes', *Westminster Review*, vol V (January 1826), pp101–130

Ayres, Robert U. (1978) *Resources, Environment and Economics: Applications of the Material/Energy Balance Principle*, John Wiley and Sons, New York

Ayres, Robert U. and van den Bergh, Jeroen C. J. M. (2005) 'A theory of economic growth with material/energy resources and dematerialization: Interaction of three growth mechanisms', *Ecological Economics*, vol 55, no 1, pp96–118

Ayres, Robert U. and Warr, Benjamin (2005) 'Accounting for growth: The role of physical work', *Structural Change and Economic Dynamics*, vol 16, pp181–20

Babbage, Charles (1832) *On the Economy of Machinery and Manufactures*, Knight, London

Barnett, Harold J. and Morse, Chandler (1963) *Scarcity and Growth: The Economics of Natural Resource Availability*, Resources for the Future/John Hopkins, Baltimore, MD

Bartlett, Albert A. (1994) 'Reflections on sustainability, population growth and the environment', *Population and Environment*, vol 16, no 1, pp5–35

Bellamy, Edward (1887 [1917]) *Looking Backward: 2000–1887*, Riverside Press, Cambridge, MA

Berkhout, Peter, Muskens, Jos and Velthuijsen, Jan (2000) 'Defining the rebound effect', *Energy Policy*, vol 28, nos 6 and 7, pp425–432

Berg, Maxine (1980) *The Machinery Question and the Making of Political Economy*, CUP, Cambridge, UK

Besiot, W. and Noorman, K. J. (1999) 'Energy requirements of household consumption', *Ecological Economics*, vol 28, pp367–383

Binswanger, Hans Christof (2006) *Die Wachstumsspirale*, Metropolis, Marburg, Germany

Binswanger, Mathias (2001) 'Technological progress and sustainable development: What about the rebound effect?' *Ecological Economics*, vol 36, no 1, pp119–132

Birol, Fatih and Keppler, Jan Horst (2000) 'Prices, technological development and the rebound effect', *Energy Policy*, vol 28, no 6/7, pp457–469

Boulding, Kenneth E. (1966) 'The economics of the coming spaceship earth', in Henry Jarrett (ed) *Environmental Quality in a Growing Economy*, Johns Hopkins, Baltimore, MD

Brewer, Anthony (1991) 'Economic growth and technical change: John Rae's critique of Adam Smith', *History of Political Economy*, vol 23, no 1, pp1–11

Brookes, Leonard (1978) 'Energy policy, the energy price fallacy and the role of nuclear energy in the UK', *Energy Policy*, vol 6, no 1, pp94–106

Brookes, Leonard (1979) 'A low energy strategy for the UK', *Atom*, vol 269 (March), pp73–78

Brookes, Leonard (1990) 'The greenhouse effect: The fallacies in the energy efficiency solution', *Energy Policy*, vol 18, no 2, pp199–201

Brookes, Leonard (2000) 'Energy efficiency fallacies revisited', *Energy Policy*, vol 28, nos 6 and 7, pp355–366

Burniaux, Jean-Marc, Martin, John P., Oliveira-Martins, Joaquim and van der Mensbrugghe, Dominique (1995) 'Carbon abatement, transfers and energy efficiency', in Ian Goldin and L. Alan Winters (eds) *The Economics of Sustainable Development*, CUP, Cambridge, UK

Cantillon, Richard (1755 [1931]) *Abhandlung über die Natur des Handels im allgemeinen*, translation by Hella Hayek, introduction by Friedrich A. Hayek, Verlag von Gustav Fischer, Jena, Germany

CEPE (Centre for Energy Policy and Economics) (2003) *Begleitende Evaluation der Wirkungsanalyse 2002 von EnergySchweiz*, SwissEnergy, Bern, Germany

Cipolla, Carlo M. (1962 [1974]) *The Economic History of World Population*, 6th edition, Penguin, Harmondsworth, UK

Clapp, Brian William (1994) *An Environmental History of Britain since the Industrial Revolution*, Longman, London and New York

Cleveland, Cutler, Costanza, Robert, Hall, Charles A.S. and Kaufmann, Robert (1984) 'Energy and the U.S. economy: A biophysical perspective', *Science*, vol 225, pp890–897

Cleveland, Cutler J. and Ruth, Matthias (1998) 'Indicators of dematerialization and the materials intensity of use', *Journal of Industrial Ecology*, vol 2, no 3, pp15–50

Costanza, Robert (1980) 'Embodied energy and economic evalutation', *Science*, vol 210, pp1219–24

Dahlström, Kristina and Ekins, Paul (2006) 'Combining economic and environmental dimensions: Value chain analysis of UK iron and steel flows', *Ecological Economics*, vol 58, no 3, pp507–519

Dahmus, Jeffrey B. and Gutowski, Timothy G. (2005) 'Efficiency and production: Historical trends for seven industrial sectors', working paper, US Society for Ecological Economics Conference, July 2005, Takoma, Washington

Daly, Herman E. (1992) 'Allocation, distribution, and scale: Towards an economics that is efficient, just, and sustainable', *Ecological Economics*, vol 6, pp185–193

Daly, Herman E. and Cobb, Clifford W. (1989) *For the Common Good*, Beacon, Boston, MA

De Bruyn, Sander and Opschoor, John B. (1997) 'Developments in the throughput–income relationship: Theoretical and empirical observations', *Ecological Economics*, vol 20, no 3, pp255–268

Defra (UK Department of the Environment, Food, and Rural Affairs) (2002) Energy Efficiency Commitment, www.defra.gov.uk/environment

Domar, Evsey D. (1962) 'On total productivity and all that', *Journal of Political Economy*, vol 70, no 5, pp597–608

DTI (UK Department of Trade and Industry) (2006) 'The energy challenge: Energy review report', July 2006, www.berr.gov.uk/files/file31890.pdf

Etzioni, Amitai (1998) 'Voluntary simplicity: Characterization, select psychological implications and societal consequences', *Journal of Economic Psychology*, vol 19, pp619–643

EEB (European Environmental Bureau) (2000) Document 2000/021, Sarah Keay-Bright, www.eeb.org/publication/general.htm

EnergieSchweiz (2007) 'Wirkungsanalyse', www.bfe.admin.ch/energie/00588/

Fawcett, Tina (2004) 'Carbon rationing and personal energy use', *Energy and Environment*, vol 15, no 6, pp1067–1083

Fouquet, Roger and Pearson, Peter J. G. (2006) 'Seven centuries of energy services', *Energy Journal*, vol 27, no 1, pp139–177

Giampietro, Mario (1994) 'Sustainability and technological development in agriculture', *BioScience*, vol 44, no 19, pp677–689

Giampietro, Mario and Mayumi, Kozo (2000) 'Multiple-scale integrated assessments of social metabolism: Integrating biophysical and economic representations across scales', *Population and Environment*, vol 22, no 2, pp155–210

Greenberg, Dolores (1990) 'Energy, power and perceptions of social change', *American Historical Review*, vol 95, no 3, pp693–714

Greenhalgh, Geoffrey (1990) 'Energy conservation policies', *Energy Policy*, vol 18, no 4, pp293–299

Greening, Lorna A., Greene, David L. and Difiglio, Carmen (2000) 'Energy efficiency and consumption – The rebound effect – A survey', *Energy Policy*, vol 28, nos 6 and 7, pp389–401

Grubb, Michael J. (1990a) *Energy Policies and the Greenhouse Effect, Vol. 1*, Dartmouth Publishing Co., Aldershot, UK

Grubb, Michael J. (1990b) 'Communication: Energy efficiency and economic fallacies', *Energy Policy*, vol 18, no 8, pp783–785

Hannon, Bruce (1975) 'Energy conservation and the consumer', *Science*, vol 189, pp95–102

Hannon, Bruce (1998) 'Letter to the Editor', *Ecological Economics*, vol 27, no 2, pp215–216

Hearn, William Edward (1864) *Plutology: Or the Theory of the Efforts to Satisfy Human Wants*, Macmillan and George Robertson, London and Melbourne

Herring, Horace (2006) 'Energy efficiency – A critical view', *Energy*, vol 31, pp10–20

Hinterberger, Friedrich, Luks, Fred and Schmidt-Bleeck, Friedrich (1997) 'Material flows vs. "natural capital": What makes an economy sustainable?', *Ecological Economics*, vol 23, no 1, pp1–14

Hotelling, Harold (1931) 'The economics of exhaustible resources', *Journal of Political Economy*, vol 39, no 2, pp137–175

Howarth, Richard B. (1997) 'Energy efficiency and economic growth', *Contemporary Economic Policy*, vol XV, no 4, pp1–9

INFRAS (2003) 'Begleitende Evaluation der Wirkungsanalyse 2002 von EnergieSchweiz', Energie Schweiz, Bern

Jänicke, Martin, Mönch, Harald, Ranneberg, Thomas and Simonis, Udo E. (1989) 'Structural change and environmental impact', *Intereconomics*, vol 24 (Jan–Feb), pp24–35

Jevons, William Stanley (1865 [1965]) *The Coal Question*, 3rd edition, Augustus M. Kelley, New York

Jevons, William Stanley (1871 [1911]) *The Theory of Political Economy*, 5th edition, Augustus M. Kelley, New York

Jones, Richard (1831) *Essay on the Distribution of Wealth and the Sources of Taxation*, Murray, London

Kaufmann, Robert (1992) 'A biophysical analysis of the energy/real GDP ratio: Implications for substitution and technical change', *Ecological Economics*, vol 6, pp35–56

Khazzoom, J. Daniel (1980) 'Economic implications of mandated efficiency in standards for household appliances', *Energy Journal*, vol 1, no 4, pp21–40

Lantz, V. and Feng, Q. (2006) 'Assessing income, population and technology impacts on CO_2 emissions in Canada: Where's the EKC?' *Ecological Economics*, vol 57, no 2, pp229–238

Lauderdale, Earl of (1804) *An Inquiry into the Nature and Origin of Public Wealth and into the Means and Causes of Its Increase*, Arch. Constable and Co., Edinburgh

Levett, Roger (2004) 'Quality of life eco-efficiency', *Energy and Environment*, vol 15, no 6, pp1015–1026

Lovins, Amory B. (1988) 'Energy saving from more efficient appliances: Another view', *Energy Journal*, vol 9, pp155–162

Luzzati, Tomasso and Orsini, Marco (2007) 'Natural environment and economic growth: Looking for the energy-EKC', in Sergio Ulgiati (ed) *Proceedings of the 5th Biennial Workshop in Advances in Energy Studies*, Porto Venere, Italy, 12–16 September 2006, ORT and VERLAG

Malthus, Thomas Robert (1798 [1976]) *An Essay on the Principle of Population*, Philip Appleman (ed), W. W. Norton, New York

Malthus, Thomas Robert (1820 [1986]) *Principles of Political Economy*, 2nd edition, E. A. Wrigley and David Souden (eds), Pickering, London

Malthus, Thomas Robert (1824 [1986]) 'On political economy', in E. A. Wrigley and David Souden (eds) *Essays in Political Economy*, Pickering, London

Malthus, Thomas Robert (1825) *Essays on Political Economy*, edited by E. A Wrigley and David Souden. Pickering, London

Manne, A. S. and Richels, R. G. (1990) 'CO_2 emission limits: An economic cost analysis for the USA', *Energy Journal*, vol 11, no 2, pp51–74

Martinez-Alier, Juan (1987) *Ecological Economics: Energy, Environment and Society*, Blackwell, Oxford

Marx, Karl (1887) *Capital: A Critical Analysis of Capitalist Production*, London/MEGA (Karl Marx Friedrich Engels Gesamtausgabe), Dietz Verlag, Berlin

McCulloch, John Ramsay (1825) *The Principles of Political Economy*, William and Charles Tait, Edinburgh and Longman and Co., London

Meadows, Donella H., Meadows, Dennis L., Randers, Jørgen and Behrens, William W. III (1972) *The Limits to Growth: A Report for the Club of Rome's Project on the Predicament of Mankind*, Potomac Associates/Earth Island, London

Mill, John Stuart (1848 [1965]) *Principles of Political Economy, with Some of their Applications to Social Philosophy*, J. M. Robson (ed), University of Toronto and Routledge and Kegan Paul, Toronto and London (1848 edition published by John W. Parker, West Strand, London)

Moezzi, Mithra (2000) 'Decoupling energy efficiency from energy conservation', *Energy and Environment*, vol 11, no 5, pp521–537

Mundella, Anthony J. (1878) 'What are the conditions on which the commercial and manufacturing supremacy of Great Britain depend, and is there any reason to think they have been, or may be, endangered?', *Journal of the Statistical Society of London*, March, pp87–126

Netting, Robert McC. (1981) *Balancing on an Alp*, CUP, Cambridge, UK

Norgard, J. S. (2006) 'Consumer efficiency in conflict with GDP growth', *Ecological Economics*, vol 57, no 1, pp15–29

NRC (National Research Council) (2002) *Effectiveness and Impact of Corporate Average Fuel Economy (CAFE) Standards*, National Academies Press, Washington, DC

Opschoor, Hans (1995) 'Ecospace and the fall and rise of throughput intensity', *Ecological Economics*, vol 15, no 2, pp137–140

Pascual, Unai (2002) 'Land use intensification potential in slash-and-burn farming through improvements in technical efficiency', *Ecological Economics*, vol 52, no 4, pp497–511

Pearce, David (1987) 'Foundations of an ecological economics', *Ecological Modelling*, vol 38, pp9–18

Perlin, John (1989) *A Forest Journey: The Role of Wood in the Development of Civilization*, Norton, New York

Petty, William (1675 [1899]) 'Political Arithmetik', in Charles Henry Hull (ed) *The Economic Writings of Sir William Petty*, CUP, Cambridge, UK

Pimentel, David, Harman, Rebecca, Pacenza, Matthew, Pecarsky, Jason and Pimentel, Marcia (1994) 'Natural resources and an optimum human population', *Population and Environment*, vol 15, no 5, pp347–369

Prettenthaler, Franz E. and Steininger, Karl W. (1999) 'From ownership to service use lifestyle: The potential of car sharing', *Ecological Economics*, vol 28, pp443–453

Price, B. B. (1998) 'Rae's theory of the history of technological change', in O. F. Hamouda, C. Lee and D. Mair (eds) *The Economics of John Rae*, Routledge, London

Princen, Thomas (1999) 'Consumption and environment: Some conceptual issues', *Ecological Economics*, vol 31, no 3, pp347–63

Radetzki, Marian and Tilton, John E. (1990) 'Conceptual and methodological issues', in John E. Tilton (ed) *World Metal Demand: Trends and Prospects*, Johns Hopkins for Resources for the Future, Baltimore, MD

Rae, John (1834 [1964]) *Statement of Some New Principles on the Subject of Political Economy, Exposing the Fallacies of the System of Free Trade, and of Some Other Doctrines Maintained in the 'Wealth of Nations'*, Augustus M. Kelley, New York

Reijnders, Lucas (1998) 'The factor X debate: Setting targets for eco-efficiency', *Journal of Industrial Ecology*, vol 2, no 1, pp13–22

Rhee, Hae-Chun and Chung, Hyun–Sik (2006) 'Change in CO_2 emission and its transmissions between Korea and Japan using international input–output analysis', *Ecological Economics*, vol 58, no 4, pp788–800

Ricardo, David (1817 [1951]) *On the Principles of Political Economy and Taxation*, 3rd edition, Pierro Sraffa (ed), CUP, Cambridge, UK

Ricardo, David (1820–22) *Notes on Malthus' Principles of Political Economy*, Piero Sraffa (ed), CUP, Cambridge, UK

Robinson, Joan (1956 [1965]) *The Accumulation of Capital*, 2nd edition, Macmillan, London

Rosenberg, Nathan (1982) *Inside the Black Box: Technology and Economics*, CUP, Cambridge

Rosenberg, Nathan (1994) *Exploring the Black Box: Technology, Economics, and History*, CUP, Cambridge

Roy, Joyashree (2000) 'The rebound effect: Some empirical evidence from India', *Energy Policy*, vol 28, nos 6–7, pp433–438

Rudin, Andrew (2000) 'Let's stop wasting energy on efficiency programs – energy conservation as a noble goal', *Energy and Environment*, vol 11, no 5, pp539–551

Saint-Paul, Gilles (1995) 'Discussion', in Ian Golding and L. Alan Winters (eds) *The Economics of Sustainable Development*, Cambridge University Press, Cambridge, UK

Sanne, Christer (2000) 'Dealing with environmental savings in a dynamical economy – How to stop chasing your tail in the pursuit of sustainability', *Energy Policy*, vol 28, nos 6 and 7, pp487–495

Sanne, Christer (2002) 'Willing consumers – Or locked in? Policies for a sustainable consumption', *Ecological Economics*, vol 42, pp273–287

Saunders, Harry D. (1992) 'The Khazzoom-Brookes postulate and neoclassical growth', *Energy Journal*, vol 13, no 4, pp131–148

Saunders, Harry D. (2000a) 'A view from the macro side: Rebound, backfire and Khazzoom-Brookes', *Energy Policy*, vol 28, nos 6 and 7, pp439–449

Saunders, Harry D. (2000b) 'Does predicted rebound depend on distinguishing between energy and energy services?', *Energy Policy*, vol 28, nos 6 and 7, pp497–500

Saunders, Harry D. (2005) 'A calculator for energy consumption changes arising from new technologies', *Topics in Economic Analysis and Policy*, vol 5, no 1, article 15

Say, Jean-Baptiste (1803 [1836]) *A Treatise on Political Economy*, translation by Prinsep, introduction by Munir Quddus and Salim Rashid, Transaction, New Brunswick, NJ, and London

Say, Jean-Baptiste (1820 [1936]) *Letters to Malthus*, translation by Harold J. Laski, George Harding's Bookshop Ltd., London

Schipper, Lee and Meyers, Stephen (1992) *Energy Efficiency and Human Activity*, CUP, New York

Schipper, Lee, Haas, R. and Sheinbaum, C. (1996) 'Recent trends in residential energy use in OECD countries and their impact on carbon dioxide emissions: A comparative analysis of the period 1973–1992', *Mitigation and Adaptation Strategies for Global Change*, vol 1, no 2, pp167–196

Schipper, Lee and Grubb, Michael (2000) 'On the rebound? Feedbacks between energy intensities and energy uses in IEA countries', *Energy Policy*, vol 28, nos 6 and 7, pp367–388

Schmidt-Bleeck, Friedrich (1994) *Wieviel Umwelt braucht der Mensch? MIPS – Das Mass für Ökologisches Wirtschaften*, Birkhäuser, Berlin

Schmookler, Jacob (1966) *Invention and Economic Growth*, Harvard University Press, Cambridge, MA

Schor, Juliet (1992) *The Overworked American: The Unexpected Decline of Leisure*, Basic Books, New York

Schor, Juliet (1999) *The Overspent American: Upscaling, Downshifting and the New Consumer*, Basic Books, New York

Schumpeter, Joseph A. (1912 [1926]) *Theorie der wirtschaftlichen Entwicklung*, 2nd edition, Duncker und Humblot, Leipzig, Germany

Schurr, Sam H. and Netschart, Bruce C. (1960) *Energy in the American Economy, 1850–1975*, Johns Hopkins, Baltimore, MD

Schurr, Sam (1985) 'Energy conservation and productivity growth: Can we have both?' *Energy Policy*, vol 13, no 2, pp126–132

Sieferle, Rolf Peter (2001) *The Subterranean Forest*, White Horse Press, Cambridge

Simms, Andrew (2005) *Ecological Debt. The Health of the Planet and the Wealth of Nations*, Pluto, London

Sismondi, Jean Charles Léon Simonde de (1819 [1827]) *Nouveaux Principes d'Économie Politique ou de la Richesse dan ses Rapports avec la Population*, 2 vols, 2nd edition, Delaunay, Paris

Smil, Vaclav (2003) *Energy at the Crossroads: Global Perspectives and Uncertainties*, MIT Press, Cambridge, MA

Smith, Adam (1776 [1976]) *An Inquiry into the Nature and Causes of the Wealth of Nations*, R. H. Campbell, A. S. Skinner and W. B. Todd (eds), Clarendon Press, Oxford

Solow, Robert (1957) 'Technological change and the aggregate production function', *Review of Economics and Statistics*, vol 39, pp312–20

Solow, Robert (1970) *Growth Theory: An Exposition*, Clarenden Press, Oxford

Sorrell, Steve and Dimitropoulos, John (2006) 'The rebound effect: Microeconomic definitions, extensions and limitations', working paper, April 2006, UKERC/SPRU, London

Spengler, Joseph J. (1959) 'John Rae on economic development: A note', *Quarterly Journal of Economics*, vol 73, pp393–406

Sraffa, Pierro (1951) 'Introduction', in Ricardo, David (1817 [1951]) *On the Principles of Political Economy and Taxation*, 3rd edition, Pierro Sraffa (ed), CUP, Cambridge, UK

SwissEnergy (2004) 'Partner for the climate: Annual Report of SwissEnergy 2003/04', Eidgenössisches Departement für Umwelt, Verkehr, Energie und Kommunikation and Bundesamt für Energie, Bern

van den Bergh, Jeroen C. J. M. (1999) 'Materials, capital, direct/indirect substitution, and mass balance production functions', *Land Economics*, vol 75, no 4, pp547–561

Veblen, Thorstein (1899 [1998]) *The Theory of the Leisure Class*, Prometheus Books, Amherst, NY

Victor, Peter A. (1991) 'Indicators of sustainable development: Some lessons from capital theory', *Ecological Economics*, vol 4, no 3, pp191–213

von Liebig, Justus (1851) *Familiar Letters on Chemistry*, Taylor, Walton and Mabely, London

von Weizsäcker, Ernst, Lovins, Amory B. and Lovins, L. Hunter (1997) *Factor Four: Doubling Wealth – Halving Resource Use*, Earthscan, London

Wackernagel, Mathis and Rees, William (1996) *Our Ecological Footprint: Reducing Human Impact on the Earth*, New Society, Gabriola Island, DC

Wackernagel, Mathis and Rees, William (1997) 'Perpetual and structural barriers to investing in natural capital: Economics from an ecological footprint perspective', *Ecological Economics*, vol 20, no 3, pp3–24

Weisz, Helga, Krausmann, Fridolin, Amann, Christof, Eisenmenger, Nina, Erb, Karl-Heinz, Hubacek, Klaus and Fischer-Kowalski, Marina (2006) 'The physical economy of the European Union: Cross-country comparison and determinants of material consumption', *Ecological Economics*, vol 58, no 4, pp676–698

Wirl, Franz (1997) *The Economics of Conservation Programs*, Kluwer Academic, Boston, MA

3

The Jevons Paradox: The Evolution of Complex Adaptive Systems and the Challenge for Scientific Analysis

INTRODUCTION: THE IMPOSSIBILITY OF DEFINING EFFICIENCY WHEN MODELLING ACROSS DIFFERENT SCALES AND DIMENSIONS OF ANALYSIS

The question of whether or not an increase in energy efficiency leads to the promotion of energy saving has been debated since the 1973 OPEC oil embargo. Many environmentalists suggest that improving the efficiency of energy use is an effective policy instrument to reduce global CO_2 emissions. On the other hand, the opposite view (the so-called 'Khazzoom–Brookes postulate') maintains that an increase in energy efficiency, as characterized at the microeconomic level, can 'backfire', leading to an increase in energy use, at the macroeconomic level, rather than to a reduction (Brookes, 1979; Khazzoom, 1980; Herring, 1999; Saunders, 2000) – a detailed discussion of this issue has been given by Alcott in Chapter 2 of this book.

The problem is made difficult by the fact that an empirical investigation of the relationship between improvements in energy efficiency and the rebound effect has to face three conceptual problems yet to be fully explored. The first of these is how to define and measure energy efficiency when dealing with complex adaptive systems[1] operating on multiple tasks across different hierarchical levels and scales. For example, an individual human being uses different energy inputs for different goals, which can only be defined on different timescales. These goals could be getting the daily meals, building a house to live in, providing an education for the children or contributing to the development of the person's cultural heritage. If we want to calculate the efficiency at which an individual human being uses 'energy' or other resources for achieving all these goals, then we have to use different variables, which can only be defined at different hierarchical levels of analysis, requiring the adoption of different temporal scales. This makes it impossible to obtain with a simple calculation an assessment of an overall unified efficiency measure for this diversified set of tasks. The second conceptual problem is how to distinguish changes in energy efficiency which are due to a change in technological coefficients (when the system performs 'the same set of transformations' but 'better') from changes in energy efficiency due to a change in the profile of tasks to be performed (when the system finds more convenient methods to perform 'something else' instead of the original set of transformations). This is the case of price-induced substitution when considering energy consumption at the macroeconomic level. And the third conceptual problem is how to separate the effect of changes due to extensive variables – for example increase in population – from the effect of changes due to intensive variables[2] – for example improvement in energy efficiency. For example, at a given point in time a given country can adopt cars which are twice as efficient as the old ones. However, if at the same time this country experiences a dramatic increase in population entailing three times more circulating cars, then the increase in population will totally offset this efficiency improvement. In this case, overall data of energy consumption referring to the country as a whole have to be disaggregated and analysed in relation to the combination of extensive variables (number of circulating cars) and intensive variables (average mileage of the fleet).

The purpose of this chapter is to provide a different perspective on the discussion about the link between increases in efficiency and sustainability. That is, we want to discuss this link in relation to an evolutionary interpretation of the Jevons Paradox. Adopting this perspective leads us to deal with two different issues. The first of these is the epistemological challenge posed by evolution to quantitative analysis: living systems when evolving in time have the peculiar ability to 'become something else' (changing both structures and functions) while remaining 'the same' (by preserving their individuality). And the second is the thermodynamic analysis of the challenge posed by evolution to quantitative analysis: two contrasting thermodynamic principles provide a sort of yin–yang tension for evolving systems. An increase in efficiency (doing better what we want to do now) makes it possible to allocate a larger fraction of the available resources in adaptability (learning how to do different things). But to increase efficiency now, one has to eliminate obsolete solutions from the existing portfolio – what we used to do in the past. For example, technological progress in agriculture has eliminated, in developing countries, animal power from the farms. This implies that improvements in efficiency within a given context (in this case farming in the oil era) do imply a reduction of adaptability in the long term (if we will run out of oil). Moreover, investments in adaptability represent a cost for society (in the short term) – for example a large fraction of the resources invested in R&D do not provide useful results. Yet investing in adaptability is the only option available in the search for new solutions providing higher efficiency when the boundary conditions change (in the long term).

Therefore, when analysing the Jevons Paradox from an evolutionary perspective, we can say that the idea that 'an increase in energy efficiency always promotes sustainability' is very simplistic. When dealing with complex adaptive systems operating across multiple scales, an alternative approach is required for analysing their performance in relation to sustainability. That is, when representing and analysing evolving metabolic systems organized in nested hierarchies, innovative theoretical frameworks are needed that can properly take care of the analysis of circular causations – in other words chicken and egg paradoxes – and multiple scales. This requires going beyond the paradigm of reductionism.[3]

This chapter is organized as follows: the first section below provides two useful narratives about the Jevons Paradox; we then introduce a few theoretical issues needed to better understand the nature of the epistemological challenge associated with the analysis of this paradox. These theoretical issues address the peculiar organization of living systems organized across hierarchical levels over multiple scales. In particular the concepts of holons and holarchies and the need to make a distinction between intensive and extensive variables are two crucial concepts to be discussed. Addressing this set of theoretical issues is crucial in order to understand the systemic failure in predicting future behavioural patterns of evolving systems. These theoretical issues imply, when dealing with evolutionary trajectories, that it is impossible to use the concept of efficiency for planning the best course of action. The subsequent section explores the nature of the Jevons Paradox using the thermodynamic narrative and defines two different concepts of efficiency. These two definitions of efficiency refer to two non-equivalent choices of a time horizon in which the stability of a given pattern of metabolism can be assessed. From this perspective we can see that the Jevons Paradox reflects the existence of a natural tension between two contrasting principles (the minimum entropy production and the maximum energy flux).[4] These seemingly contrasting principles refer to different representations of the process of evolution. This analysis of contrasting principles, combined with the set of concepts and epistemic tools in the preceding section, provides a theoretical resolution of the paradox. That is, the Jevons Paradox just reflects natural patterns associated with evolution, which entails contrasting goals in relation to different objectives, which can only be defined at different hierarchical levels and scales. It may appear as a paradox simply because those using conventional scientific analytical tools are often confused when forced to deal with the perception and representation of the process of evolution. A final section provides the conclusions of this elaborated analysis.

THE JEVONS PARADOX REVISITED: TWO USEFUL NARRATIVES

The myth of the dematerialization of developed economies: Are elephants dematerialized versions of mice?

As noted earlier, a sound analysis of the changes induced by technological improvement should explicitly address the different effects of intensive and extensive changes. If this is not done properly, then it becomes easy to be misled by the counter-intuitive behaviour of evolving complex systems. The myth of dematerialization of developed economies can be used as a good practical example of the systemic errors which can be generated by the use of intensive variables for dealing with the analysis of changes in socioeconomic systems. It is to expose such a systemic error that the dematerialization of developed economies is discussed here in relation to the metabolism of elephants and mice.

An economic definition of energy efficiency is based on the calculation of the so-called economic energy intensity (EEI) number. This number reflects the ratio between MJ of energy consumed by the economy (the biophysical input calculated in energy terms) and the GDP produced by the economy (the resulting economic output, calculated in terms of added value measured in a given currency referred to a given year). This is then used to study changes in the evolution of socioeconomic systems. By adopting this approach, however, one can get the false impression that technological progress is decreasing the dependence of modern economies on energy. For example, when looking at EEI and GDP per capita (GDP_{PC}) in the US economy, many seem to be reassured by the notion that technological progress has been associated with a decreasing EEI and an increasing GDP_{PC}. Indeed, the historic series between 1950 and 2005 (Figure 3.1) confirms the steady decrease in EEI and the persistent increase in GDP_{PC}, supporting the neoclassical economic view. However, this interpretation of 'improvement' in economic terms reflects the choice of using data that only refer to intensive variables (EEI and GDP_{PC}). But these intensive variables are not necessarily useful for checking the compatibility of the socioeconomic process

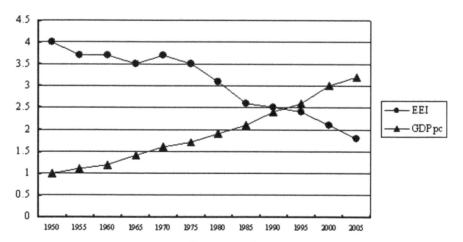

Figure 3.1   *Economic energy efficiency and GDP per capita in the US economy*

Source: compiled from the data in Heston et al (2006) and Annual Energy Review (2006)

with ecological processes, in other words for dealing with sustainability issues. Herman Daly (1996) provides an effective critical metaphor for the systemic error represented by this approach:

> *The physical exchanges crossing the boundary between the total ecological system and the economic subsystem constitute the subject matter of environmental macroeconomics. These flows are considered in terms of their scale or total volume relative to the ecosystem, not in terms of the price of one component of the total flow relative to them and the subsystem, rather than the pricing and allocation of each part of the total flow within the human economy or even within the nonhuman part of the ecosystem. ... Optimal allocation of a given scale of resource flow within the economy is one thing (a microeconomic problem). Optimal scale of the whole economy relative to the ecosystem is an entirely different problem (a macroeconomic problem). The micro allocation problem is analogous to allocating optimally a given amount of weight in a boat. But once the best relative location of weight has been determined, there is still the question of the absolute amount of weight the boat should carry. This absolute optimal scale of load is recognized in the maritime institution of the Plimsoll line. When the water level hits the Plimsoll line the boat is full, it has reached its safe carrying capacity. Of course, if the weight is badly allocated, the water line will touch the Plimsoll line sooner. But eventually, as the absolute load is increased, the water will reach the Plimsoll line even for a boat whose load is optimally allocated. Optimally loaded boats will still sink under too much weight, even though they may sink optimally!* (Daly, 1996, p48–50).

Regarding the size of the society in relation to the size of the environmental services available to a country, let us now consider the relative changes in two extensive variables that also describe the changes that took place in the US between 1950 and 2005. These two extensive variables are total energy consumption (TEC) and population. The ratio of these two variables (TEC/population) provides another intensive variable (energy consumption per capita or EC_{PC}). By considering these two extensive variables, the picture of the changes taking place in the US economy between 1950 and 2005 is dramatically changed. The historical data on TEC, population and EC_{PC} over the same period of time are given in Figure 3.2. The extensive variable population steadily increased over the period. The intensive variable EC_{PC} seems to have attained a sort of a saturation point starting around the year 1970.[5] However, when looking at the overall movements of TEC (reflecting the combined changes of EC_{PC} and population), there is no evidence of dematerialization of the economy. The US has experienced steady increases in both the size of humans (in other words

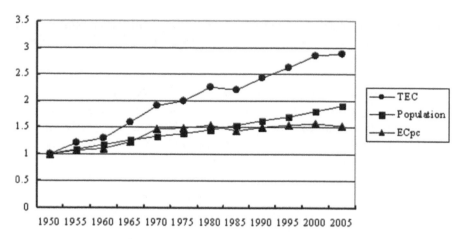

Figure 3.2   *Total energy consumption, population and energy consumption per capita in the US economy*

Source: compiled from the data in Heston et al (2006) and Annual Energy Review (2006)

human activity driven by the metabolism of endosomatic energy) and machines (activity generated by exosomatic devices driven by the metabolism of exosomatic energy), in terms of Georgescu-Roegen's bioeconomic viewpoint.[6]

Finally, the degree of dematerialization induced by technological progress in the US economy can be checked by simultaneously analysing the two views provided by Figure 3.1 and Figure 3.2. Over the considered period, the energy efficiency of the economy more than doubled (a reduction of EEI, the energy consumed per unit of GDP), which had the effect of increasing the aggregate consumption of commercial energy in the US economy by almost three times! As indicated by Figure 3.2, the aggregate energy consumption of the US increased not only because of an increase in consumption per capita but also because of an increase in population size. This latter phenomenon is explained not only by differences between fertility and mortality, but also by immigration, driven by the attractive economy, since strong gradients in the standard of living among countries – generated by gradients in efficiency – tend to drive labour from poorer to richer countries (Giampietro, 1998):

> *For example, the dramatic improvement in energy efficiency that the state of California has achieved in the past decade (in terms of the intensive variable useful energy/energy input) will not necessarily curb total energy consumption in that state. Present and future technological improvement are likely to be nullified by the dramatic increase in immigration, both from outside and inside the US, which makes the Californian population among the fastest growing in the world. Again, we find the systematic failure of accounting for the change in boundary*

conditions induced by the change in technology at the root of this counterintuitive trend. (Giampietro, 2003, p12)

The same systemic error is evident when comparing the performance of developed and developing countries. For example:

... in 1991 the US operated at a much better value of EEI than China (12.03MJ/$ versus 69.82MJ/$ respectively). On the other hand, because of this greater efficiency the US managed to have a GNP per capita much higher than in China (22,356$/year versus 364$/year respectively). That is, if we change the mechanisms of mapping changes moving to an extensive variable (by multiplying the energy consumption per unit of GNP by the GNP per capita) the picture is totally reversed. In spite of the significantly higher economic energy efficiency, the energy consumed per US citizen is 11 times higher than that consumed by a Chinese citizen. (Giampietro, 2003, p11).

Note that in this quote GNP rather than GDP is used in the calculation of EEI (this does not affect the validity of the discussion).

In relation to the myth of the dematerialization of developed economies, it is time to mention the striking similarity in the pattern of relating the two variables – 'intensity of metabolism' and 'size' – found when comparing socioeconomic systems and biological organisms. In biology it is well known that animals with a smaller body size have a higher rate of energetic metabolism per kg of biomass. There is an abundant literature on this phenomenon: a good overview for empirical analyses is Peters (1986); more recent theoretical applications are described in Brown and West (2000). Using available data (organized in tables or in parameters for equations that can be applied to different 'typologies' of animals), we can calculate, for example, the relationship between size and metabolic intensity for mammals of different size, as shown in Figure 3.3. For example, a male mouse weighing 20g – an extensive variable – has a metabolic rate of 0.06W (joule/second), yielding for male mice a metabolic rate of 3W/kg of body mass – an intensive variable. In stark contrast, a male elephant weighing 6000kg has a total metabolic rate of 2820W yielding a metabolic rate of 0.5W/kg of body mass, or six times less that of the male mouse (Peters, 1986, p31).

If in this case we apply the same reasoning used by some neoclassical economists – using an intensive variable assessing EEI to describe the process of dematerialization of modern economies – we would find a bizarre result. Looking at animal biomass across the evolutionary ranking and using an intensive variable assessing the energy expenditures per unit of biomass, we would find a quite peculiar way of defining a process of 'dematerialization' in mammals. Since 10,000kg of elephant consumes 4700W whereas 10,000kg of mouse consumes 30,000W, we have to conclude that elephants, with their low energy intensity

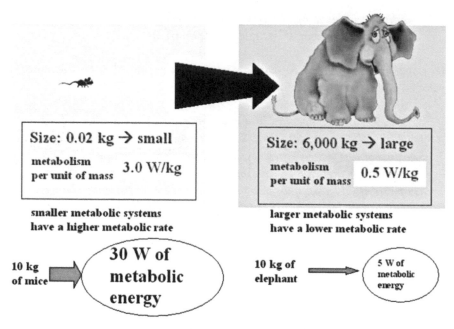

Figure 3.3   *Are elephants dematerialized versions of mice?*

per unit of biomass, should be considered *a 'dematerialized' version of mice*, with a high energy intensity per unit of biomass, as shown in Figure 3.3. After all, this is exactly what we are told by neoclassical economists. According to this perception, the process through which very poor countries (based on location-specific subsistence economies) are evolving into large developed countries (based on the pattern of production and consumption typical of the global economy) is described as a process of 'dematerialization' of the world economy!

The Jevons Paradox and the evolution of cars

The backfire effect in the energy analysis literature was closely examined by none other than William Stanley Jevons, one of the four founders of neoclassical economics. The Jevons Paradox[7] (Jevons, 1990; Giampietro and Mayumi, 1998; Mayumi et al, 1998; Clark and Foster, 2001; Alcott, 2005; Polimeni and Polimeni, 2006) was first enunciated by Jevons in his 1865 book *The Coal Question* (Jevons, 1865) (a detailed historical analysis is given in Chapter 2). Briefly, it states that an increase in output/input ratio – the 'efficiency' in using a resource – leads, in the medium/long term, to increased use of that resource rather than to a reduction. At that time, Jevons was discussing possible trends of future consumption of coal and reacting to scenarios advocated by technological optimists who ignored increasing yearly rates of consumption. As is happening today, some contemporaries of Jevons believed humans could rely on dramatically increasing the 'economy' of steam engines in order to reduce coal consumption. In the face of such a claim, Jevons correctly indicated that more efficient engines would increase coal consumption in established uses as well as expanding the possible uses of coal for human activities. Therefore increases in efficiency would boost the rate of consumption of existing coal reserves rather than reducing it.

The Jevons Paradox seems to be true not only with regard to demand for coal and other fossil energy resources but also with regard to demand for resources in general. Doubling the efficiency of food production per hectare over the last 50 years (the Green Revolution) did not solve the problem of hunger. Unfortunately, this doubling of efficiency actually made the food shortage problem worse, since it increased the number of people requiring food, the fraction of animal products in the diet and the absolute number of the malnourished (Giampietro, 1994). In the same way, doubling the number of roads did not solve the problem of traffic, but rather made the traffic condition worse since it encouraged more use of personal vehicles (Newman, 1991). As more energy-efficient automobiles were developed as a consequence of rising oil prices, American car owners increased their leisure driving (Cherfas, 1991). Along with the expected performance of cars, the number of miles driven increased; moreover, US citizens are increasingly driving heavier vehicles like minivans, pick-up trucks and four-wheel drives. And as a further example, refrigerators have become more efficient but also bigger (Khazzoom, 1987). A promotion of energy efficiency at the micro level of economic agents tends to increase energy consumption at the macro level of the whole society (Herring, 1999). In economic terms, we can describe these processes as increases in supply boosting demand in the long term, a much stronger phenomenon than Say's Law of Markets.[8]

> *The Jevons Paradox has different names and different applications: 'rebound effect' in energy literature and 'paradox of prevention' in relation to public health. In the latter case, the paradox consists of the fact that the amount of money 'saved' by prevention of a few targeted diseases leads to a dramatic increase in the overall bill of the health sector in the long term. Due to the fact that humans sooner or later must die (which is a fact that seems to be ignored by 'ceteris paribus' efficiency analysts), any increase in the lifespan of a population directly results in an increase in healthcare expenses. Besides the higher proportion of retired people in the population who need more healthcare, it is well known that the hospitalization of the elderly is much more expensive than the hospitalization of young adults.*
> (Giampietro, 2003, p7)

This last example leads us to the heart of the paradox. Technological improvements in the efficiency of a process represent improvements in *intensive variables*, defined as an 'improvement' per unit of something and under the *ceteris paribus* hypothesis (that everything else remains the same). However, an increase in efficiency would lead to resource saving only if the process of evolution did not modify the existing portfolio of behaviours in response to efficiency improvements. As a matter of fact, evolving metabolic systems, especially human systems, tend to adapt quickly and effectively to any changes in efficiency improvements. As soon as a series of 'technological improvements' are introduced into a social system, more room is generated for a further expansion of current levels of activity (for example people make more use of their old cars) within the original option space and an expansion of the option space with the addition of new possible categories and activities (for example new models of car including new features such as air-conditioning, faster acceleration or more space per person).

In more abstract terms, the first expansion refers to a quantitative change in *extensive variables within a given formal identity assigned to an observed system under analysis.* By formal identity we mean a set of attributes in relation to an observed system, a set of proxy variables and their relationships in the model representing how the observed system is supposed to behave. In this case, the dimension of the process gets bigger within the same option space and the original formal identity of the system (the same old car with more driving). The second type of expansion refers to *emergence due to a qualitative change* that requires an introduction of new concepts, categories and variables in the formal representation of a new category (a new type of car capable of expressing more functions and having different gadgets).[9] In this new category, 'car' represents an increase in the diversity of possible options within the set of accessible states for consumers looking for a vehicle. The expansion of the option space adds 'new meanings' to the original repertoire of meanings associated with the word 'car'. The introduction of these new meanings can be viewed as the emergence of new couplings between external referents – for example a sort of essence and its association, what we have in mind when we think of a car – and formal identities in the representation of the modelled system.

The distinction between intensive and extensive variables has important implications when it comes to modelling changes in efficiency. A formal model can handle the quantification of changes only by keeping the same set of attributes, the same set of proxy variables (intensive and extensive) and their functional relationships associated with the formal identity of the modelled system. Therefore, within the given model, the handling of quantitative changes requires only an update of the value taken by the given set of selected variables. Unfortunately, qualitative changes cannot be handled by using the same old formal identity of the system under investigation. If the model of a car evolves into something different, the modeller must add new attributes to obtain a new quantitative characterization of the modelled system – the evolution of Fiat models of a utility car are shown in Figure 3.4. When dealing with a model of a car which has much better amenities than the original model (four-wheel drive, air-conditioning, much more power or much more payload, for example), it is

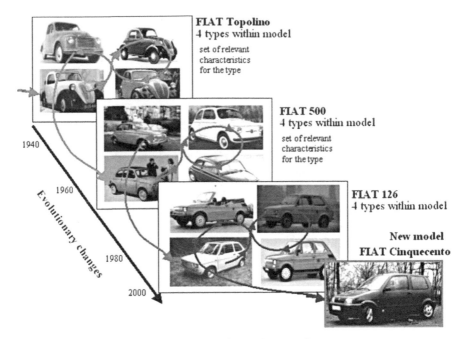

Figure 3.4   *The evolution of cars*

meaningless to compare the two cars' performances only in terms of gas consumption.

Looking at the evolution of cars in time, we can say that the introduction of more efficient car engines has determined that some features – such as air-conditioning – which were optional in the past became standard features of modern cars. Thus an increase in efficiency in one of the attributes of performance – generating power in the engine – has led to the addition of a new set of standard attributes in the definition of 'what modern cars are and should be' from a consumers' perspective. Increases in efficiency have also made it possible to introduce new categories of cars such as minivans or SUVs. This represents an increase in the diversity of possible options within the set of accessible states for consumers looking for a vehicle. This expansion of the option space has added 'new meanings' to the original set of meanings associated with the word 'car'. The introduction of these new meanings can be viewed as the *emergence* of new couplings between external referents (what we have in mind when we think of a car) and formal identities used in the mathematical representation adopted in the models (the syntactic representation of the car). Because of this, the set of variables and attributes useful to provide a quantitative representation of the performance of a SUV will differ from those useful to provide a quantitative representation of the performance of a small car.

Put another way, it is possible to have a bifurcation with regard to the interpretation and prediction of the effect of an increase in efficiency when dealing with the process of evolution. This bifurcation can be characterized as follows. First, if we assume that the formal identity used in the model will remain relevant and useful in the future – which implies the validity of the *ceteris paribus* assumption for both the context of the observed and the context of the observer – then we can say that increasing the efficiency over the given set will lead to a decrease in energy consumption. On this side of the logical bifurcation we are operating under the *ceteris paribus* assumption when making predictions about the effect of an increasing efficiency. And second, if we assume that the formal identity to be used in models will change over time, then we must admit that there is a systemic problem with the interpretation of the meaning of the word 'car'. On this side of the logical bifurcation the set of attributes and epistemic categories required to capture potential relevant features when modelling cars is open and expanding. This can be due to the fact that the observed car model evolves over time, or that the expectations of the consumers (determining the attributes to be observed) evolve in time. A strategy of increasing the efficiency of the various functions associated with a given formal identity of a car – an existing model – will provide a window of opportunity to add new features and attributes to future cars and thereby new meanings for potential buyers. In other words, the ability to produce more efficient cars will unavoidably expand the current expected performance of a car and therefore lead to the production of different cars. These new models of cars will require the use of different formal identities, with new categories and new proxy variables, in order to reflect the addition of new functions and behaviours to the new models. We are, therefore, on this side of the logical bifurcation, operating under the assumption that the phenomenon of 'emergence' is something unavoidable when dealing with evolution. Emergence requires the use of new concepts, categories and variables in the formal representation (in quantitative models). Because of emergence, the original formal identity used in a model loses its validity and will have to be replaced by another formal identity. Therefore the phenomenon of emergence makes it impossible to predict the effect of an increase in efficiency, while still using the original formalization of the concept of efficiency.

The important point for our discussion is that 'how will the system expand?' and 'what are the consequences of this expansion?' are questions which cannot be answered by those studying the system from within the original model, based on the adoption of the old formal identity. These two questions cannot be answered by making inferences based on the given set of attributes under the *ceteris paribus* assumption. Put another way, when increasing the efficiency of a given process based on a given set of formal models – when we 'try to save the world' by making a given process more efficient – we are unintentionally increasing the likelihood of emergence. This implies a sort of Catch-22 for the usefulness of formal models:[10] the more these are useful for increasing improvements in efficiency, the quicker the status quo will change and the more likely it is that these formal models will become useless for making long-term predictions. This is why, when dealing with the analysis of evolution, it is crucial to adopt complementing views about change: a steady-state view, which makes it possible to deal with concepts such as efficiency, better design and reliability in the short run, combined with an evolutionary view, which makes it possible to deal with alternative useful concepts such as adaptability, diversity and uncertainty in the long run.

THEORETICAL ISSUES IN RELATION TO THE JEVONS PARADOX: EXPLAINING THE FAILURE OF REDUCTIONISM WHEN DEALING WITH THE ISSUE OF EVOLUTION

An overview of the epistemological impasse

The scientific representation and analysis of the evolution of complex adaptive systems pose a series of serious epistemological challenges. This is due to the peculiar set of characteristics of metabolic systems which are organized in nested hierarchical levels and have the ability to evolve simultaneously across different scales and to learn. These epistemological challenges, faced when dealing with the evolution of social or ecological systems, are not encountered in the traditional fields of application of reductionist science. In fact, social and ecological systems are necessarily:

1   open systems which cannot be in thermodynamic equilibrium; this entails that they must be exchanging matter and energy with the environment on which they depend for their establishing structures and functions[11] (Prigogine, 1961; Glansdorff and Prigogine, 1971; Nicolis and Prigogine, 1977);

2   hierarchically organized and operating on multiple spatial-temporal scales (Allen and Starr, 1982; O'Neill, 1989; Ahl and Allen, 1996) – thus the characteristic structural and behavioural patterns and changes expressed at different hierarchical levels are evolving at different temporal paces (Giampietro, 2003);

3   autopoietic systems (Maturana and Varela, 1980 and 1998) – *poiesis* literally means creation or production, and autopoietic indicates 'the circular organization' of living systems and the dynamics of the autonomy proper to them as a unity (Maturana and Varela, 1980); and

4   organized in a particular type of nested hierarchy based on the concept of 'holon' (Koestler, 1967, 1969 and 1978). This concept will be illustrated in more detail below.

Below we elaborate briefly on the implications of these four key characteristics.

*Complex adaptive systems are dissipative systems
and therefore 'becoming' systems*
Because of their first characteristic, namely their metabolic nature, social and ecological systems are always qualitatively, as well as quantitatively, evolving or co-evolving with their environment. According to the vivid image proposed by Prigogine (1987), the problem of modelling those systems is associated with the fact that they are always 'becoming' something else. This characteristic makes a *substantive* formal representation of their changes basically impossible.[12] Due to this unavoidable evolutionary nature, arithmomorphic models[13] (differential equations and other conventional inferential systems) – the standard tool of hard science – are far from satisfactory for representing and simulating their evolution.

The phenomena of emergence, as discussed in the evolution of cars, points at the obvious, but often neglected, fact that a metabolic system must be necessarily a 'becoming system' and therefore requires a continuous update of the selection of attributes together with proxy variables and their relationships – the formal identity assigned to the observed system – used to describe its behaviour. As a matter of fact, a model *itself*, even if successfully used in the past, does not evolve in time in terms of a selection of variables and equations, whereas the modelled system undergoes continuous qualitative changes. Therefore, once the attributes selected for the formal identity of the observed system become no longer relevant for predicting behaviours of the system, the proxy variables and their analytical relations must be automatically discarded. Then a new set of attributes with a new set of proxy variables and relations should be introduced. After these selections are made, both a new formal identity (a given and finite set of relevant attributes which can be represented using a given and finite set of proxy variables) and a new inferential system (a finite set of axioms, rules and algorithms) must be introduced. Therefore, when adopting any formal model for the representation of the behaviour of an evolving metabolic system, the need of dealing with these two categories of changes – a change in formal identity and a change in inferential system – is always with us. This is the reason why we should accept an unavoidable dose of uncertainty and ignorance in relation to representation of an evolving metabolic system. Any change in the formal representation entails the need to introduce new measurement schemes, new data set collection methods for dealing with a different definition of initial conditions and boundary conditions, and a new time interval for the simulation of behaviours of the observed system.

Complex adaptive systems are organized across hierarchical levels and scales
The second epistemological problem is generated by the fact that living evolving systems are hierarchically organized and operate across multiple spatial-temporal scales. This implies that alternative yet perfectly legitimate methods of description can coexist when studying the evolution of a living system (Whyte et al, 1969)

and that the characteristic structural and behavioural patterns expressed at different hierarchical levels are evolving in time, but at different paces (Giampietro, 2003). This can explain why alternative (and also contrasting) assessments of efficiency can be found when considering simultaneously tasks referring to different temporal horizons. Here we provide an example of contrasting scientific statements that are determined by the pre-analytical decision of using non-equivalent descriptions about the same observed system.

Building on the insightful arguments provided by Mandelbrot (1967), we provide a trivial example of ambiguity associated with the interpretation of a given formal statement. The coastline of Maine in the US can be perceived and represented as oriented toward the east, south, west or north (Figure 3.5a). In this example, the epistemological ambiguity generating the contrasting statements is associated with the interpretation of the label 'geographic orientation of the coast'. At the level of the continent (level $n+1$), Maine is on the East Coast of the US. In this case we can consider Maine as the level n and North America as its context, as level $n+1$. However, at the level of the county, defined at the hierarchical level $n-1$, the coast of Maine, as a state, faces toward the south. Moving again towards a lower hierarchical level – the town in the chosen county at the hierarchical

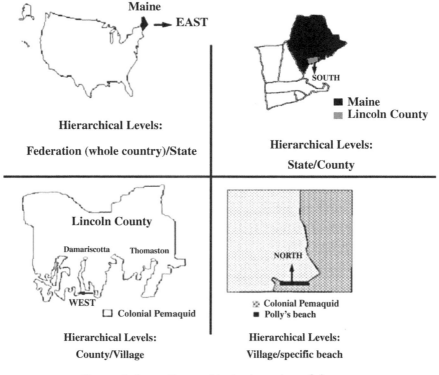

Figure 3.5a *Geographical orientation of the coast*

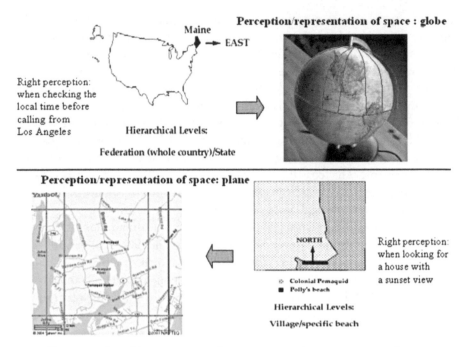

Figure 3.5b   *Two non-equivalent perceptions/representations of space*

level *n–2* – some parts of the coast of Maine face the west. Moving further downward to another lower level, we find an individual beach which is facing north (in the figure it is Polly's Beach in the town of Pemaquid). In all these cases rigorous experiments can be conducted to prove the truth of each orientation. However, contrasting statements will remain because of the initial epistemological ambiguity in the definition of 'the geographical orientation of the coast', which can be legitimately perceived and therefore represented at different scales in non-equivalent ways. As a matter of fact, the original example of Mandelbrot (1967) – 'How long is the coast of Britain?' – was exactly focused on the impossibility of formalizing, using a given metric, objects which are defined simultaneously in different ways across different scales. In his seminal paper he made the point that a numerical assessment of the length of the coast of Britain depends not only on the characteristics of the observed object, but also on the choice of the scale of the map used for representing it. The choice of how to observe a fractal object will affect the relative representation and quantification.

When defining Maine as being on the east coast of North America, we are adopting the relative position of continents on the globe as an external referent for assigning meaning to the expression 'geographic orientation of the coast'. In this narrative, continents are the relevant parts of the whole, and their relative positions are defined over a sphere. This is illustrated in the upper part of

Figure 3.5b. On the other hand, when defining Polly's Beach as oriented toward the north, we are adopting the direction indicated by a compass as an external referent when standing on the beach. Such an orientation of the local beach comes from a representation of the area as related to a flat map. This is illustrated in the lower part of Figure 3.5b. Within such a narrative the lines going to the North Pole are represented in the relative formalization as parallel. Now, it is possible to relate the representation based on a flat map (lower part of Figure 3.5b) to the relative position of different flat maps on the globe, however, the formalizations of the reciprocal relation in space of the lines going to the North Pole are non-equivalent and not reducible when carried out at these different scales.

If we want to make a phone call from Los Angeles to Maine and we want to calculate the difference in time zones, then it is the narrative of the relative position of continents – the east coast with meridians converging at the North Pole – that provides the useful analysis. Within this narrative the indication given by a compass operated on the beach is not only useless, but it will be misleading. Whereas if we want to buy a house on Polly's Beach with the porch facing the sunset, then for pertinent analysis we need the narrative of Polly's Beach facing north on a flat map where the meridians will never converge.

The main point we want to make with this example is that the quantification of concepts referring to complex systems operating across different scales – a class to which metabolic systems belong by default – is never substantive and always dependent on a procedural agreement on how to perceive and represent the system, in a given context in relation to a given purpose.

By analogy, the example above illustrates how it is possible to find non-equivalent definitions and measurements of efficiency. In turn, these non-equivalent definitions of efficiency can lead to contrasting assessment of its effects on each component as well as the whole system. These non-equivalent assessments are unavoidable when simultaneously considering tasks referring to different spatial-temporal perspectives, as will be discussed.

Autopoiesis entails impredicativity (handling chicken and egg paradoxes)
The third epistemological predicament is associated with the circular causality typical of the evolution of living systems. This epistemological predicament has been systematically ignored by traditional scientific treatment. In fact, accepting as a given fact the existence of circular causality in life would entail a major epistemological problem for reductionism. It would imply that focusing on just a linear causality and a single scale at a time will always provide partial analysis of a given state of affairs. That is, the various elements generating an autopoietic process can only be perceived and represented by adopting a set of different spatial-temporal scales, meaning that the relative process is not formalizable in substantive terms as done by reductionism. In fact, the perception and

representation of a direction of causation between two events is determined by the choice of attributes and scales made when choosing a given model. A few familiar examples – discussed by Giampietro (2003, Chapter 3) – referring to this predicament are listed below:

1   The notion of consumer democracy – that is, the idea that, based on the purchases they make, consumers determine what goods and services are produced. At the large scale, consumers reign over the economy. At the local scale, on the other hand, consumers can only choose among products that have been already produced – those that are already available on the market.

2   The number of predators affects the number of prey, when looking at this relationship on a particular time horizon. But we could find a reverse causal relation – the number of prey affecting the number of predators – when looking at the same prey–predator relationship on a different time horizon. This process of impredicativity has been clearly proved in quantitative terms in ecology (Carpenter and Kitchell, 1987).

3   In democratic countries, governments and parliaments affect the behaviour of individual citizens, while citizens affect government and parliament behaviour over a time horizon which includes a few elections.

The predicament of impredicativity is more difficult to accept since it goes deep against the simplifications associated with reductionism. On the other hand, living systems heavily depend on impredicativity for their self-organization (Rosen, 2000). This characteristic is crucial to explain their ability to express 'autopoiesis'. Impredicativity has to do with the familiar concept of the 'chicken and egg logic paradox' – you need to know about an existing chicken in order to be able to recognize an egg as such, but at the same time you need an egg to have a chicken in the first place. Bertrand Russell called the predicament associated with impredicativity a vicious circle (quoted in Rosen, 2000, p90)[14] and for this reason it has always been avoided by conventional formal analysis (Kleene, 1952; Lietz and Streicher, 2002).

However, it should be noticed that even in theoretical physics, one of the most conservative scientific fields in terms of reductionism, we are now witnessing the acknowledgement of such a concept in, for example, superstring theory. A Nobel Prize winner in physics, Gell-Mann (1994) made a clear reference to the bootstrap principle (based on the old saw about the man that could pull himself up by his own bootstraps) by describing the concept as follows:

> ... the particles, if assumed to exist, produce forces binding them to one another; the resulting bound states are the same particles, and they are the same as the ones carrying the forces. Such a particle system, if it exists, gives rise to itself. (Gell-Mann, 1994, p128)

This passage basically means that as soon as the various elements of a self-entailing process – defined in parallel on different levels – are all present and at work, then such a process will become able to stabilize itself. This process of autopoiesis will then arrive at a point where it generates a predictable and detectable entity which can be distinct from noise by an observer that has learned about it.

To conclude we can say that impredicativity requires considering events taking place simultaneously on different scales and that therefore require the use of different 'choices for perceiving and representing time' – in other words time differentials and time horizons. This translates into the need of using non-equivalent models which are not-reducible, incommensurable, logically independent or, as stated in physical jargon, 'incoherent' (Giampietro et al, 2006a). Put another way, they would require using simultaneously different timescales for perceiving and representing different directions of causality. Exactly because of this peculiar epistemological challenge, impredicative loops are out of the reach of conventional analytical tools.

On the contrary, conventional analytical tools are developed within a paradigm assuming that all the phenomena of reality can be explained by adopting a linear definition of cause and effect. In technical jargon we can say that conventional analytical tools are assuming that all the phenomena of the reality can be described within the same descriptive domain (referring to the same substantive definition of space and time) and by using a set of reducible models (Rosen, 2000). This assumption excludes the possibility of having two or more narratives about the same event, which are logically independent and therefore providing different explanations. That is, reductionism assumes that if two social actors have contrasting perspectives about what is relevant in a given situation, one of the two must have bad models. Put another way, given a situation to be faced, all contrasting perspectives about what is relevant and what should be done but one are wrong.

However, contrasting but legitimate perspectives about what is relevant and useful in an analysis are not only possible, but also are necessary to preserve diversity. In relation to the Jevons Paradox this last epistemological challenge can be used to resolve the apparent contradiction between non-equivalent definitions of efficiency and thermodynamic principles, for example the minimum entropy generation and the maximum energy flux. As will be discussed below, a successful surviving trajectory of evolution must result in the ability to establish an impredicative loop between increases in efficiency, an attribute very relevant in the short run, and an increase in adaptability, an attribute very relevant in the long run. If only one of the two strategies is adopted – increasing adaptability by reducing efficiency or increasing efficiency by reducing adaptability – a negative side-effect will show up either in the short or the long term. If the two types of strategies are effectively combined and reinforced, a harmonious autopoietic process across multiple scales will come out.

Introducing the concept of the holon and its elusive identity
The last characteristics of complex adaptive systems – the peculiar organization of nested hierarchies made up of 'holons' – is less known to the general public and will be presented in detail in the rest of this section. The concept of the holon was proposed by Arthur Koestler (1967, 1969 and 1978) as an epistemic tool useful for handling the complexity of living systems. Koestler's holon is a combination of two Greek words: *holos*, meaning the whole with constraints from the macroscopic view, and the suffix *on*, referring to a part or particle (such as a proton or neutron) with its constraints from the microscopic view. Thus a holon has a double nature of 'whole' and 'part', which is also a typical feature of the components of metabolic systems such as social and ecological systems. The metabolic systems in their structural and functional aspects can be represented as a hierarchy of self-regulating nested holons. This hierarchy of holons was termed a holarchy (Koestler, 1967). The part/whole dualities must be able to express a valid identity in relation to both structural and functional terms (see also Allen and Starr, 1982, pp8–16).[15] Therefore, the term holon may be associated with the duality typical of the Eastern concept of yin–yang. Holons must be simultaneously perceived and represented in relation to two aspects determined by:

1   how/what (a realization for a structural type[16] – for example either an airplane or a balloon); and
2   why/what (a realization for a functional type – for example something capable of carrying objects in the air).

The how/what view addresses the local-scale view in order to be able to define a pertinent structural type and its fabrication – in other words how parts behave *within* the whole. On the other hand, the why/what view entails adopting the large-scale view in order to define a relevant functional type and its behaviour – in other words how the whole behaves *within* its associative context.

In the next section we provide an analysis, based on the concept of holons and holarchies, of two major epistemological problems faced when attempting quantitative analysis of the evolution of socioeconomic or ecological systems, arguing that it is impossible to have a substantive perception and representation of a holon since it refers to a perception of a successful (and ambiguous) realization of a structural and functional type in a given context and that it is impossible to have a substantive one-to-one mapping between structures and functions in holons belonging to evolving holarchies. The continuous emergence of new structural types, functional types or successful functional combinations makes it impossible to select a formal identity for a holon which will remain valid into the future.

The special challenge faced when analysing holons and their evolution

The impossibility of fully formalizing the representation of a holon

Consider the President of the United States as an example of holon. In this example the considered holon refers to a natural identity in a social system, 'the US President'. In relation to this natural identity, Mr George W. Bush is the actual 'realization of an organized structure' or the 'incumbent' in the 'function' or 'role' of President of the US.[17] Different individual human beings (a realization of the same structural type) can perform such a role for a limited finite time (examples given on the right of Figure 3.6). By contrast, the role of the US presidency, as a social function, is a functional type which has a time horizon estimated in the order of centuries associated with a given set of symbols and encoded information for its representation (examples on the left of Figure 3.6). Therefore, the perception and representation of individual realizations of either the structural type (the various incumbents) and/or the functional type (the various images and written definitions associated with the institutional role of US presidency) can only be obtained adopting a different selection of relevant attributes (see Figure 3.6). Even so, when we refer to the 'US President', we loosely address such a holon, without making a clear distinction between the role (functional type) and the incumbent (structural type) performing that role. As a matter of fact, you cannot have an operational US President without the joint

The **role** (function) is written in the US constitution and preserved by institutions with a life span of centuries

The **incumbent** (structure) is associated with a particular individual and has a turnover time of a 8 years maximum

Figure 3.6   *The two sides of the holon: The President of the US*

existence of a valid role associated with an effective structural type and a valid incumbent verified in the election process. The role needs institutional settings for its validity. The effective structural type should be someone born in the US who is 35 years of age or older and has been a legal resident in the US for 14 years according to the US Constitution (Article II, Paragraph 1).

The ambiguity associated with the concept of holon can be explained using the metaphor of the three blind men touching the elephant. That is, an effective coupling between a valid role and a valid incumbent is logically independent on the duality between a type and an individual instance of the type. Any valid incumbent represents just one of the possible realizations of the required incumbent type. The identity of Mr Bush as a particular realization of the organized structural type (an adult human being) able to perform the specified function of the 'US President' is logically independent from the identity of the role of the US President. That is, the shared and commensurate images related to the physiological characteristics of human beings can be used for describing Mr Bush or another US citizen as a possible realization of this structural type. On the other hand, the shared and commensurate images related to the typical characteristics of the social institution, the US President, cannot capture some of the special characteristics of Mr Bush the human being. This is a matter of scale and the set of relevant attributes selected for defining the equivalence class determining the given type. Human beings were present in America well before the writing of the US constitution. In the same way, the American constitution has a lifespan much longer than any of the incumbents in the role of the US President.

The impossibility of fully formalizing the representation of the evolution of holons and holarchies due to the unavoidable phenomenon of emergence

Two examples of the impossibility of fully formalizing the representation of the evolution of holons and holarchies due to the unavoidable phenomenon of emergence are given in Figures 3.7a and 3.7b, which explore the different facets of a timepiece. The examples given in Figure 3.7a illustrate many different structural types (many 'how/whats') that map onto the same functional type (the same 'why/what'). In this case, after defining the performance associated with a given role, we can learn how to increase the efficiency of structural types. That is, we can compare the performance of the various how/whats (structural types) against the given set of expected behaviours associated with the why/what (functional type).

The inverse situation is shown in the examples given in Figure 3.7b, in which we have the same how/what, the structural type realized in a given physical object, but many why/whats. When an individual realization of the timepiece is moved to a different context, the same organizational structure of this individual physical object can map onto a different why/what. The same

same function - keeping time → many structural types (templates/organized structures)

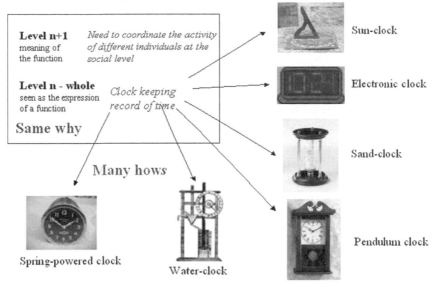

Level n+1
meaning of
the function

*Need to coordinate the activity
of different individuals at the
social level*

Level n - whole
seen as the expression
of a function

*Clock keeping
record of time*

Same why

Many hows

Sun-clock

Electronic clock

Sand-clock

Pendulum clock

Spring-powered clock

Water-clock

Figure 3.7a   *The same why corresponds to many hows*

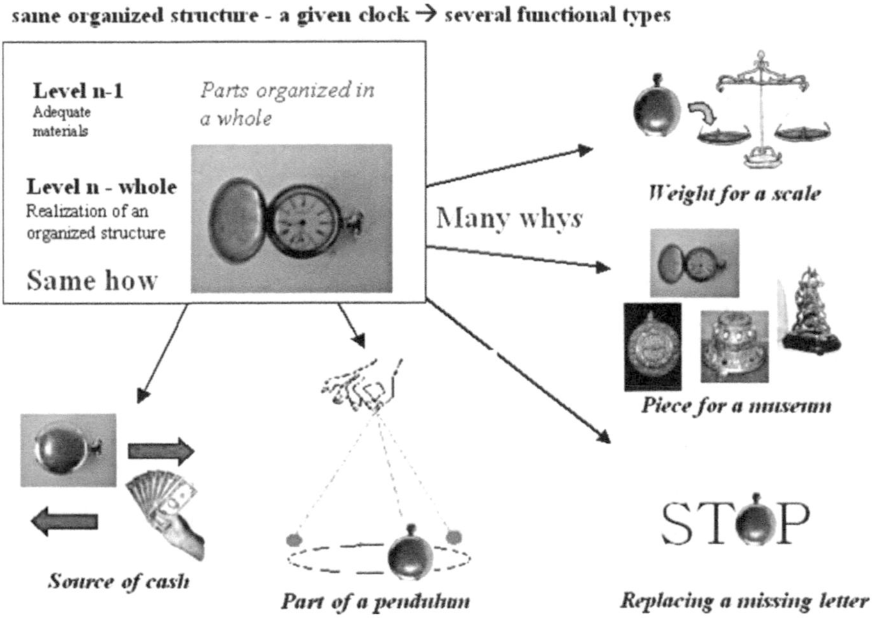

same organized structure - a given clock → several functional types

Level n-1
Adequate
materials

*Parts organized in
a whole*

Level n - whole
Realization of an
organized structure

Same how

Many whys

Weight for a scale

Piece for a museum

Source of cash

Part of a pendulum

STOP

Replacing a missing letter

Figure 3.7b   *The same how corresponds to many whys*

object can play different functions. That is, depending on the context, the same realization of a given structural type can perform many different roles. As a matter of fact, George Bush also plays several different roles in his private life, besides that of the US President. In the examples given in Figure 3.7b, a realization of the structural type 'old mechanical clock' can become the structural type as 'object worth putting in a museum' or 'source of cash' or 'weight to be used on a scale'. The first one of these new functional types is associated with the shared feeling of a society for the need to preserve records and a common memory of their process of learning how to keep time. This is an example of emergence, in which an existing structural organization (carried out by an individual realization) is coupled to a different associative context (a latent demand for new functions expressed by the system of knowledge in which meanings are created and preserved).

This ability to generate novelty is determined by the existence of two dualities:

1   a duality between structural and functional types, which have to be wisely coupled to get an operational whole, in order to be able to share commensurate experience; and

2   a duality between 'individual realization' and 'expected type', which makes it possible for the observer to perceive the 'expected' patterns associated with the type to which the particular realization is supposed to belong, either structural or functional.

The ambiguity associated with these dualities (especially the first one) determines the possibility of bifurcations associated with potential changes in the goals of the observer/story-teller.

So far, we have not emphasized enough the role of the observer/story-teller in creating new functions. Changes in the goals of the observer/story-teller play an important role in the emergence of new functions. In fact, a given story-teller/agent/observer can assign different meanings to the same object, within the local priorities given to the perception/representation of its own interaction with a relevant reality. This is possible because any given realization of a given structural type – even a realization based on a design such as a timepiece, which has been expressly made with the goal of keeping time – can always be given a new meaning by a story-teller facing an alternative task. The more pressing the task, the easier it becomes to find new meanings in terms of new functions for existing entities. A few examples of this situation are given in Figure 3.8. Depending on the circumstances, the same physical entity – the timepiece made for keeping time – can be used for obtaining much needed privacy, or for making available drugs to a drug addict. The story-teller/agent/observer can associate a given physical entity with functions which are usually associated with other structural types when looking for a function to be expressed.

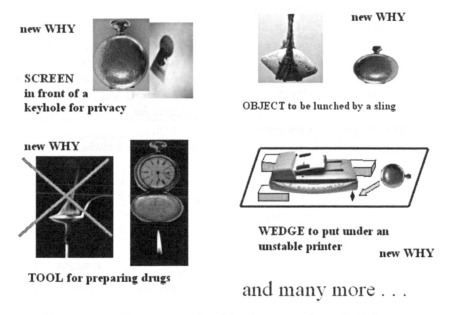

new WHY

SCREEN
in front of a
keyhole for privacy

new WHY

OBJECT to be lunched by a sling

new WHY

TOOL for preparing drugs

WEDGE to put under an
unstable printer

new WHY

and many more . . .

Figure 3.8   *Emergence induced by changes in the goals of observers*

When dealing with the evolution of holarchies, we should expect a continuous loss of a one-to-one mapping between the realization of structural types and functional types. This is different from the situation found in a relationship written in a formal statement (for example in the instructions written in a computer program). There a particular symbol can be associated univocally with a given operation to be executed. But there are other situations in which, after assuming the definition of the functional type used for the model as valid – for example that humans need a device for keeping time during a day in order to organize their daily activities – it becomes possible to find several what/hows for it. That is, many structural types can map onto the same functional type (many how/whats for the same why/what of a clock, as in Figure 3.7a). When operating in this situation, the different performances of these different structural types can be compared in relation to the formalization of the function. *Here we are in the realm of efficiency based on design.* Looking at the perceived functional and structural types, it is possible to learn how to improve their efficiency based on design.

On the other hand, there are also situations in which it becomes possible for a given realization of a structural type to perform a function which is different from that for which that original structural type was originally fabricated. This can depend on sudden changes in the goals of story-teller/designer or associative context (boundary and/or initial conditions). In this case, a new useful function can push for the introduction of a new formal identity to be associated with the original system, due to a new definition of role that has to be fulfilled (Figure 3.7b).

Whenever a new natural identity is expanded to a point where it becomes a recognized essence – because of the matching of the structural and functional type with the expected associative context – a new functional type or essence is born. It should be noted that by using terms as such 'holons' and 'essences' we are blurring ontology and epistemology (for more see Giampietro et al, 2006a). An essence can be associated with the commensurate experience about a meaningful external referent, which makes it possible for an observer or a story-teller to define a useful identity for a relevant functional type. As illustrated in Figure 3.8, depending on the circumstances, very many why/whats (functional types) can be assigned to the same how/what (structural type). *This is the realm of emergence.* Emergence can be associated with the introduction of a new essence in the universe of discourse. It has to do with the ability to share the meaning associated with the name of a holon within a knowledge system.[18] This implies the existence of a recognized functional role at a large scale that is worthy of receiving a name.

As soon as a new essence is born, it becomes possible to define the relative formal identity of such a functional type by representing in analytical terms the relevant behaviour associated with the function. This makes it possible to characterize and associate the expected role with an associative context. That is, it becomes possible to formalize such a role in terms of an expected behaviour associated with expected boundary conditions.

It is only at this point, after this formalization, that it becomes possible to define the concept of 'efficiency' for this new functional type. That is, a definition of the role in analytical terms makes it easier to look for improved structural types, after ranking them in terms of their relative efficiency.

Evolution of a holarchy can be seen as a continuous process in which the definition of a functional type is used to learn how to design more efficient organized structures and the realizations of more efficient structural types makes it possible to discover new relevant behaviours and meanings beyond the original definition of function. Increasing the very low efficiency of the first steam engine made it possible to move them out of coal mines. After reaching a certain level of efficiency, steam engines were finally able to power a railway train. This was possible only when the weight of the fuel and the engine to be carried on the train was small enough compared with the power delivered. This process of evolution implies, however, that the old formal identity of the complex whole (functional type and structural type) will become obsolete and a set of new formal identities will have to be added to the original set every time a new function is acknowledged in the form of a new essence. As discussed above, increasing the efficiency of cars made it possible to expand their original definition of performance. New behaviours of cars were realized and soon associated with the meaning of the word 'car'. Again ambiguity is generated by the difficulty of catching the moving target associated with the semantics of a definition using just a quantitative ratio. In the same way, more efficient light bulbs change the definition of what is an acceptable level of lighting for rooms or outdoor spaces (both in terms of intensity and the period of time). Old cars (when considered as types of organized structures) were just expected 'to move people or goods around without using draft animals' (the original definition of the relative functional type). Modern cars are expected 'to move people around very quickly, safely and with air-conditioning' (the new definition of the relative functional type).

In conclusion, when dealing with evolution it becomes impossible to maintain over time a valid formalization of the performance of a given essence based on a given coupling of the formal identities referring to either a structural and/or a functional type. *This is the realm of ignorance faced by modellers asked to deal with evolution and emergence.*

Another look at the problem faced when attempting a simultaneous formalization of both structural and functional types is given in Figures 3.9 and 3.10. The figures show an example of three realizations belonging to two equivalence classes, which require the use of logically independent definitions for the relative types. Two different structural types – airplanes and balloons – can perform the same function, whereas two realizations of the same template may not necessarily be associated with the same functional type – flying airplanes versus toy airplanes. As a matter of fact, when we perceive a holon (the epistemic tool we use to make sense of what we perceive), we are perceiving a realization of two types (a structural type and a functional type) realized in the same physical entity. The perception and representation of one of the two types using formal

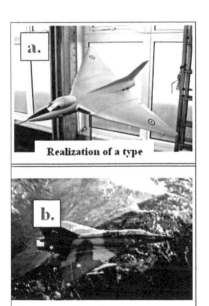

Structural Type :
expected shape and relative size of wings and fuselage

$$a. = b. \neq c.$$

Functional Type:
ability to carry loads in the air

$$b. = c. \neq a.$$

Realization of a type   Realization of a type

Realization of a type

Figure 3.9   *Three realizations of structural and functional types*

identities implies assuming a tacit agreement about commensurate knowledge of the complementing type, either functional or structural. That is, when looking at the various pictures in Figure 3.9 and perceiving a jet airplane among the objects shown, our perception is based on the signals generated by a realization of the structural type (the pattern shown in the picture). However, this ability requires the presence of the expected pattern of the structural type in our mind (the shape of a jet airplane recognized in the picture, which is associated with the knowledge about airplanes in our knowledge system). In turn, this pre-existing knowledge about airplanes indicated that airplanes perform a functional type which is relevant for our knowledge system. This is why we know about the existence of jet airplanes in the first place.

The crucial point of this discussion is that any perception is not just based on the coupling of two types (a structural and a functional type) but depends on (1) the interaction with an individual instance of either a functional and/or a structural type and (2) the existence of the relative essence (the agreement about the relevance of the external referent associated to a functional type) in the knowledge system.

What is interesting in the example of different aspects of flying objects in Figure 3.10 is the difference in the formal representation adopted in science for structural and functional types. Templates referring to structural organization can be handled using images – in terms of expected geometric/spatial relations over parts and wholes – which are *not changing in time, since all structural types are*

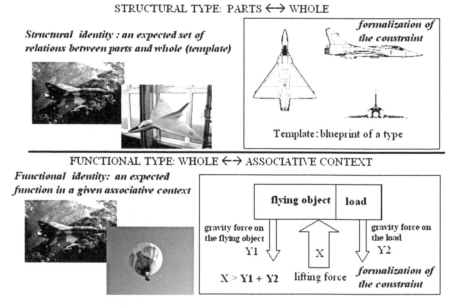

Figure 3.10   *Differences between structural identity and functional identity*

rate-independent. On the other hand, the representation of functions (in technical jargon an expected behaviour within an associative context, which must be an admissible environment for the holon) tends to be handled using differential equations. In this way scientific analysis can provide a representation of a given mechanism of causality described as the ability to induce *expected changes over a given state space, since functional types are time-dependent*. This formal representation of change, however, is possible only because of a pre-analytical definition of a given bound and finite universe of possible changes (events) to which the equations apply. This requires that the state space, the quantized universe of possible changes, which is associated with the representation provided by the equations, is given and *cannot change in time*.

The epistemological conundrum associated with the evolution of holarchies: Acknowledging the concept of complex time

Building on the insight provided by the work of Robert Rosen (1977, 1985 and 2000), Giampietro (2003) proposes the concept of complex time. The concept of complex time was originally introduced to deal with the existence of distinct timescales (time differentials and time durations) to be considered when studying the evolution of complex adaptive systems across nested hierarchical levels. Perceiving, representing and simulating the evolution of these systems, both in quantitative and qualitative analyses, requires the simultaneous use of different formalizations of 'time' and 'change'. However, the concept of complex time was further extended to effectively deal with the role of observers/story-tellers in determining the relevance of the narratives selected in the process of issue definition and problem structuring. When dealing with the perception and representation of the process of evolution, both the observer and the observed are captured in time, as is the knowledge system within which the observation takes place. In this situation it is unavoidable that the definition of relevance provided by the cultural context about what and how to observe the external world will change in time. Therefore the concept of complex time plays a vital role in examining how our knowledge system evolves. In particular there are at least four types of discrete time intervals which have to be individuated to characterize the evolution in time of a perception/characterization of changes (Giampietro, 2003; Giampietro et al, 2006a):

1   The concept of Δt is related to the pace of perceived changes of the observed system within a given representation. This discrete time interval is used to represent changes within the equations of the models. For example, a set of differential equations based on a given Δt produces a representation of a given behaviour in this framework. For each Δt it is necessary to give the starting time and the duration of simulation. Each Δt corresponds to a particular

measurement scheme to be adopted for dealing with the experiment. A selected formal identity and an inferential system do not change for the duration of the simulation.

2   The concept of $\Delta\tau$ is related to the pace of perceived changes of what is observed within a given narrative. This discrete time interval is important because it determines when an obsolete formal identity and inferential system must be replaced by a new one. This problem arises because of the *becoming* of the observed reality, requiring the use of new categories to represent relevant attributes for its representation. The example here is of the changes in US cars in the last 50 years (see again the example in Figure 3.4) in which air-conditioning became a standard attribute of such an identity.

3   The concept of $\Delta\theta$ is related to the pace of changes in the interests of the observer/story-teller within the universe of the available narratives. Changes in the interests of the observer/story-teller require an update in the selection of narrative and formal identities assigned to the observed system in the observation protocol. This is the set of discrete time intervals at which the selection of relevant narratives used in the model becomes obsolete. This problem is generated by the *becoming* of the observer/analyst/society around the scientist, which requires a continuous updating of issue definition and problem structuring. This may occur, for example, when the access of a given car to restricted urban areas depends on the type of pollutants emitted by its engine. This will determine the adoption of new attributes not considered before (a new narrative about cars).

4   The concept of ΔT is related to the pace of changes in the characteristics of the system of knowledge within which the process of observation and validation of observations are generated in terms of relevance based on shared meaning. This is the time interval at which autopoietic systems must redefine the right formalization of their own identity, if these systems are to survive the perturbations while preserving the meaning assigned to them. The identity of an autopoietic system defines the overall context in which the scientific activity takes place and determines the choice of priorities and modalities of how to do science in the first place. This identity is about what should be kept alive and what should be discarded when becoming something else while keeping the same perceived individuality.

To stress the peculiar meaning of the last one of the four discrete time intervals – ΔT – we can recall here the heart-wrenching line at the end of the autobiographical *Plenty-Coups: Chief of the Crows* (Plenty-Coups, 2002), referring to the history of US after the buffalos 'went away': 'And after this nothing happened …'

Whenever there is no longer a relevant story-teller to observe processes occurring in the external world, it really does not matter whether or not the reality exists ontologically in the first place, external to the story-teller's concerns and observations – then *nothing happens.*[19]

An overview of the relationships among these different definitions of time within the process generating quantitative analysis within a knowledge system is given in Figure 3.11. Scientists dealing with quantitative analysis must decide in a pre-analytical phase about the semantic problem of structuring before constructing a formal model. This implies choosing a definition of relevant reality, which depends on the identity of the relevant story-teller, and then choosing useful narratives, associated with validated beliefs, for the purpose of the analysis. Within different scientific disciplines, this chain of decisions leads to a different set of final choices of descriptive domains – in other words what is to be observed, how to select the measurement scheme and the length of the period of observation. Different scientific disciplines – for example chemistry and zoology – look at different relevant realities that are populated by different holons (effective coupling of structural and functional types) interacting among themselves. These holons can be perceived only using their specific spatial and temporal referential framework.

In contrast with complex time, we can define 'simple time' as a given definition of changes associated with the choices required for representing the passing of 'time' within a given quantitative model. In simple time, changes are represented over a finite set of variables based on a given definition of a time differential (associated with the chosen measurement schemes) and duration (associated with the validity of the assumptions used in the model). It should be noted that simple time exists only within a formalized representation of change.

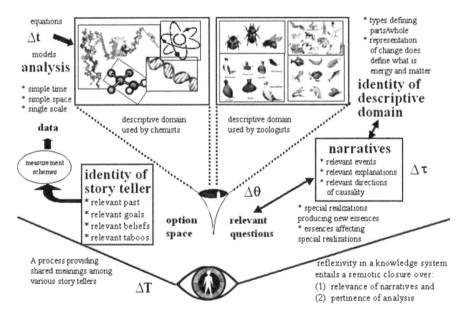

Figure 3.11   *Complex time: Four discrete time intervals*

A crucial implication of complex time is that when trying to predict changes associated with evolution, any formal model is bound to become obsolete. In fact, complex time entails that the system we want to model will become, sooner or later, something else, whereas the set of variables and inferential relation used to build the model will remain the same. For this reason, we should expect a systemic failure when using a model whose formal structure is given and not changing in time (based on a given set of types) to predict the emergence of new functions and structures in an evolving system. Here we should recall Georgescu-Roegen's severe verdict on the usefulness of econometric models to make predictions about the future:

> Even more crucial is the absence of any concern for whether the formula thus obtained will also fit other observations. It is this concern that is responsible for the success natural scientists have with their formulae. The fact that econometric models of the most refined and complex kind have generally failed to fit future data – which means that they failed to be predictive – finds a ready, yet self-defeating, excuse: history has changed the parameters. If history is so cunning, why persist in predicting it? (Georgescu-Roegen, 1976, ppxxi–xxii)

In this passage Georgescu-Roegen criticizes only the performance of econometric models in the case in which the basic formalization remains valid and the problem is with the inability to choose the right values of parameters. Thus his critique has only to do with the inability to handle the process of becoming in relation to the second time interval $\Delta \tau$.

THE NATURE OF THE JEVONS PARADOX FROM A THERMODYNAMIC PERSPECTIVE

Two non-equivalent interpretations of the concept of efficiency and the Jevons Paradox: The minimum entropy production and the maximum energy flux

As discussed at length above, the concept of efficiency always requires a semantic interpretation at the moment of formalization. For example some authors propose a distinction between *efficiency*, interpreted as a dimensionless ratio between an output and an input of a given process of conversion, which requires that both the output and the input must be measured using the same unit, and *efficacy*, interpreted as a ratio between a relevant output and a relevant input of a given process. However, this distinction is only apparently rigorous. As a matter of fact, if in a given process of conversion it is possible to make a distinction between a flow of output and a flow of input, then this distinction entails that the two flows are different in quality. This is what makes the concept of efficiency relevant. Therefore the dimensionless ratio must in any case refer to two different types (for example forms of energy) which have to be reduced to the same unit/metric. It should be noted that when handling and reducing numerical assessments of different energy forms to a single ratio, one should be extremely careful, especially when these flows are described across different hierarchical levels and scales (Giampietro, 2006; Giampietro et al, 2006a, b). In this situation it is extremely likely to get into an unavoidable ambiguity in the definition of the boundaries and the identities of the various elements involved in the process of energy transformation. When handling a set of energy transformations across different energy forms, aggregation entails a certain level of arbitrariness (Giampietro, 2006).

In relation to this unavoidable ambiguity, we want to discuss here the existence of two 'principles' proposed by different authors working in the analysis of the evolution of living systems. These two principles – the minimum entropy production and the maximum energy flux – are often interpreted as providing contrasting explanations about the thermodynamic driver of evolution. The main point of this section is that there is no contradiction between these principles; their apparent contrast just reflects different interpretations of the concept of efficiency for metabolic systems operating away from thermodynamic equilibrium.

To introduce our argument, let us start with the work of Kawamiya (1983) defining, in physical terms, two types of 'efficiency' which are relevant for humans. We use the example from Kawamiya's work since it is important to study the nature of the Jevons Paradox:

1   Efficiency of Type 1 (EFT1) refers to the ratio between output and input. However, this definition of efficiency ignores the time required to generate output. A familiar example would be the mileage obtained with a litre of fuel, which does not consider the time required for the travel.
2   Efficiency of Type 2 (EFT2) refers to the pace of generation of an output (per unit of time). This interpretation of efficiency proposed by Kawamiya ignores the amount of input required by the process to obtain output. A familiar example would be the cruising speed of a car, without considering the related fuel consumption.

Using the familiar example of the mileage of a car in relation to the cruising speed, we know the expected relationship between these two types of efficiency: if we want to increase the speed of our car above the recommended threshold for fuel-economy, we will consume more fuel per mile (Figure 3.12). In this example the concept of EFT1 refers to fuel economy (miles per gallon), which does not address the time required for travelling the miles. Therefore EFT1 would refer to an interpretation of the concept having the focus on 'energy efficiency'. The

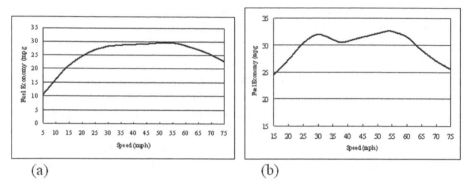

Figure 3.12   *Fuel economy and speed of cars*

Source: The United States Department of Energy (www.fueleconomy.gov) and Eartheasy.com

concept of EFT2, on the other hand, refers to the speed (miles per hour), which does not address the fuel consumption. On the contrary, EFT2 would refer to an interpretation of the concept having the focus on 'time efficiency'. Even though most drivers know that an increase in EFT2 will surely decrease EFT1, they often prefer to look for a higher EFT2 (in terms of cruising speed) rather than a higher EFT1 (in terms of less fuel consumption).

Getting into a more technical analysis of efficiency – for example the efficiency of an engine – we can say that for a real engine the value of EFT1 can vary from a minimum of zero to a maximum value 'η' of achievable efficiency, depending on the type of engine and several other relevant factors. Finally, there is an ideal engine – a Carnot engine – that can be used to study the maximum achievable efficiency in relation to thermodynamic laws. When dealing with a Carnot engine (hypothetically without friction or heat loss), we encounter a theoretical limit of efficiency (EFT1) determined by the entropy law. This value must be less than one.[20] In this situation, when the speed of a piston of an ideal Carnot engine is regarded as the output, then EFT2 is zero for practical purposes. As mentioned already, starting from this extreme condition and moving into the realm of real thermal engines, it is well known that any increase in EFT2 – an increase of the power level at which the engine can generate work – will result in a reduction of EFT1; it will increase the consumption of energy input per unit of output. The power level here represents the pace of the energy output per unit of time. Thermodynamic considerations can be used to learn the differences in logic behind the maximization of these two types of efficiency by social and economic systems.

To keep the discussion out of technical jargon, we can say that under the *ceteris paribus* condition, lower input requirements (an increase in EFT1) has beneficial effects on the stability of boundary conditions. That is, a car with higher EFT1 consumes less fuel, has to stop less frequently at service stations and emits less CO_2 – it depends less on the environment for the supply of fuel and for the sink capacity of its emissions. When considering the interaction of the whole society with its environment in relation to societal metabolism (in other words the flow of matter and energy inputs taken from the environment and the flow of wastes released into the environment), we find a similar pattern. A reduction in the physical input required by a society to perform a given activity is in and of itself ecologically benign. An increase in EFT1 decreases the rate of natural resource depletion and the stress on the environment associated with the dumping of the relative wastes. On the other hand, a higher speed of throughput, implied by an increase in EFT2, has beneficial effects on the ability of the socioeconomic system to express more complex behaviours and enlarge its domain of activities. This higher speed of the biophysical throughput is associated with a growth in the throughput of the economic process – the flow of goods and services produced and consumed in a given society according to a set of codified rules. Again a distinction would be required looking at the effect of intensive (affluence) and extensive (population) variables and the profile of distribution of the value of these variables over the population (the equity issue), but it is in general accepted that economic growth, so far, has proved to be benign to human societies. Metaphorically speaking, the ability to carry out the required task in a shorter period of time provides a real advantage over competitors to those capable of getting such an edge: time is money.

Therefore, EFT1, not EFT2, is related to the biophysical scale of the metabolism of the socioeconomic system, in other words to the amount of material throughput of a given society. When using as a metric the consumption of energy, water and other key material flows, the metabolism of socioeconomic systems can be compared to the biophysical metabolism of the ecosystems embedding them. Thus EFT1 can be regarded as a factor determining the size of the activity of the economic process compared with the size of the activity of natural processes. This entails that EFT1 should have higher priority than EFT2 as 'natural capital' becomes a limiting factor in economic growth (Daly, 1995). In thermodynamic terms maximum EFT1 would be the minimum energy throughput required to fulfil a particular structure/function in society.

The concept of EFT1 has been formalized, even though it should be regarded as a phenomenological principle rather than a physical law, by Prigogine in relation to the analysis of energy-dissipating systems using the expression 'minimum entropy production principle' (Prigogine, 1961; Glansdorff and Prigogine, 1971; Nicolis and Prigogine, 1977). The technical description of this principle states that 'linear systems obey a general inequality implying that at a *steady* non-equilibrium state, entropy production becomes a minimum compatible with the constraints applied on the system' (Nicolis and Prigogine, 1977, p45, emphasis added).[21] To formalize this kind of stability, scientists use a Liapunov function typical of control theory. It is crucial to observe that even if this principle has been developed within the field of non-equilibrium thermodynamics, its validity requires the assumption that the system under analysis must be stable in order to approach the steady state. The pattern of dissipation associated with the particular choice of variables and equations used in the model must also occur sufficiently close to equilibrium to guarantee that the formalization used remains reliable in time.

To make this discussion less technical, let us work out a practical example of a metabolic system whose formal identity remains valid because of a very strict definition of initiating and boundary conditions. This would be, for example, a cell operating within a human being, the cell being a holon, as discussed above. In this example, we have a metabolic system (the cell) operating at a local scale, which is a part of another metabolic system operating at a larger scale (a tissue within a human body). As long as the larger-scale metabolic system – the human being – remains alive, then the boundary conditions for the given cell, which received a clear identity from the instructions written in its DNA, can be assumed to remain favourable. That is, we can assume for this cell that the temperature will remain around 37 degrees Celsius, the pH of the blood will remain close to 7, and the supply of oxygen and nutrients in the blood will remain adequate, together with an adequate sink capacity for disposal of CO_2 and other chemical wastes. In this example, the combination of reliability of the initial conditions and the stability of boundary conditions expected for a human cell provides the crucial condition that the formal identity used to represent the functioning of cells will remain valid over time. That is, when dealing with metabolic systems organized in nested hierarchical levels such as social or ecological systems, we can expect that lower-level system components are fabricated according to a given blueprint and operate under a controlled set of boundary conditions (Giampietro, 2003). In this situation it is reasonable, assuming some 'natural selection' at work, to expect that structures/functions of these lower-level elements will slowly move over a trajectory determined by the principle of minimum entropy production. Learning how to reduce internal consumption is definitely positive for this type of holarchy. As discussed before, the higher the EFT1, the lower the quantity of input taken from the environment (less depletion of natural resources) and the less waste released into the environment (less environmental pollution). 'Where the supply of available energy is limited, the advantage will go to that organism which is most efficient, most economical, in applying to preservative uses such energy it captures' (Lotka, 1922, p150).

On the other hand, as discussed later on, lowering the flow of throughput (by reducing the required input to obtain the same output) generating a dissipative system implies, in the long term, lowering the diversity of options and behaviours that can be expressed by that system. This reduced option space can become a liability when boundary conditions change. As a matter of fact, expanding the ability to produce more in order to be able to consume more is a very benign solution for those living within socioeconomic systems.

This explains why, in both economic and biophysical analyses, the idea that evolution has to do with the maximization of EFT2 has been very popular. Formalizations of this principle used in economics have to do with the maximization of profit and welfare. In ecological theory, the metaphor of maximization of EFT2 has been formalized in terms of the maximization of energy flows within ecosystems. This principle has been proposed as one of the general principles of evolution for living systems: '*natural selection* tends to make the energy flux through metabolic systems a maximum, so far as to be compatible with the constraints to which the system is subject' (Lotka, 1922, p148).

Lotka's maximum energy flux principle (Lotka, 1922) has been proposed under a series of different names by several authors: 'evolution through Malthusian instability' (Layzer, 1988); 'maximum exergy degradation' (Morowitz, 1979; Jørgensen, 1992; Schneider and Kay, 1994); and the 'maximum power principle' within the analysis provided by the Odum school (Odum and Pinkerton, 1955; Odum, 1971 and 1996).

Here we would like to use a couple of metaphors to illustrate the complex interrelation between the two principles of input minimization and throughput maximization. Let us imagine the situation of people experiencing a shortage of either food or money. They can adapt to live with a very low level of food and/or money. In this way, they can successfully obtain the balancing of their dynamic budget in terms of food or money. With 'dynamic budget' we want to indicate the forced matching between the requirement and supply of metabolized flows, which can be obtained at different levels of throughput.

On the other hand, with a dynamic budget based on a low input for a low throughput they will experience a limited ability to express a diversity of behaviour; a limited domain of action in space and time; and an increased risk of collapse in case of perturbations. So an alternative strategy could be changing their pattern of activity in order to be able to get a balanced dynamic budget at a higher level of throughput, by getting more food and/or earning more money. This would require, however, changing their own identity.

In order to carry out the discussion in these terms, it should be noted that any definition of a change towards either a lower level of consumption or a higher level of consumption requires a preliminary definition of an expected standard – the given identity of the metabolic system. That is, it should be possible to define an 'identity' for the metabolic system made up of humans, which entails the existence of both biophysical and/or cultural constraints on the expected level of consumption. For example, in pre-industrial societies the consumption of grain per capita was on average around 250kg per person per year (Smil, 1994). This standard level was due to a combination of biophysical conditions, low technological levels and demographic constraints. After the industrial revolution, given a consistent excess in the supply of both food and money, humans 'learned' how to consume much more than in pre-industrial times.[22] For example, in developed societies the double conversion of grain into meat and alcoholic beverages makes it possible for people living in developed countries to consume more than 1000kg of grain per capita per year (FAOSTAT, 2007). This boosting of the level of consumption must be performed outside the human body – it would be impossible, using only physiological conversions, to directly eat such a large amount of grain. This value represents a consumption of 2.7kg of grain per capita per day (including babies, children and the elderly). That is, in order to be able to produce more grain per capita, humans had to boost their ability to consume more. But to do that, they had to include yeast generating alcohol, cows, pigs and poultry in the system of transformations, which is using grains as an input, in order to guarantee the food security of human beings, according to the chosen diet. In order to maximize their energy consumption they had to change the 'identity' of their food system. This example resonates with the discussion we had about the evolution of cars in relation to how to consume more energy – through increases in efficiency – by expanding the portfolio of functions and structures associated with the definition of their performance.

Minimum entropy production and maximum energy flux

Using the insight provided by Mandelbrot in relation to the example of apparently contrasting assessments over the orientation of the coastline of Maine, we want to make the point that the same epistemological ambiguity muddles the discussion over the minimum entropy and maximum energy flux principles. These two principles refer to different and non-equivalent perceptions and representations of evolution. Both analytical models can be useful, depending on the context and the scale; therefore, they are not in contradiction. However, care has to be taken in order to combine non-equivalent descriptions and non-reducible statements about the external world into a useful understanding of the evolutionary process. Consider, first, that the minimum entropy production principle is related to the idea of EFT1 efficiency and refers to a perception of evolution obtained from inside the system, that is on the interface between the two levels of analysis, level n (focal level – the whole) and level $n-1$ (lower level – the parts defined within the given representation of the whole). Because of this perception on the interface, the analysis of this efficiency has to be performed using a given definition of identity for the relative structure/function at a local/small scale. And second, that the maximum energy flux principle is related to the idea of the EFT2 definition of efficiency, interpreted as the ability of generating more useful work (relevant output). This interpretation of the concept of efficiency refers to a perception of evolution obtained from outside the system, seen as a black box interacting with its context. This definition refers to the interface between the two levels of analysis, level $n+1$ (the higher level or the environment, assumed to be stable) and level n (focal level – the whole). In this case, the system undergoing evolution is considered as changing its functions and structures at a large scale. Because of this perception on the interface, the types of changes taking place within the black box are simply ignored by this representation, since it does not consider what is inside the black box.

A system which is evolving according to the maximum energy flux principle will express different functions by using a series of different 'identities' of the parts/black box in relation to its local definition of structural and functional types.[23] Again the reader can recall the example of the evolution of cars in response to technological improvements.

When describing the evolution of a system on a large scale at the upper levels, the maximum energy flux principle indicates the continuous process of generation of new complexity through co-evolution. That is, this principle points at the need of increasing the mutual compatibility and adaptability of the various metabolic systems interacting within the same set of boundary conditions (for example different socioeconomic systems embedded in the same biophysical context). On a large scale, the distance from the thermodynamic equilibrium of the whole is considerably high, so the emergence of unpredictable behaviours is unavoidable and a new formal identity must be introduced to take care of change in the meaning of efficiency. 'When a dissipative structure is near such instability its entropy production reaches a relative maximum and it becomes sensitive to small fluctuations' (O'Neill at al, 1986, p105).

When describing the evolution of a system on a local scale at the lower levels, by adopting a quasi-steady-state view, we deal with dissipative components operating under a strict set of constraints within a stable set of boundary conditions (for example cells within an organism). Under these conditions, it is reasonable to assume a trend towards a continuous learning of new ways for reducing the quantities of energy and matter required to sustain the given particular function – an increase in efficiency. In this case, the minimum entropy production principle is much more relevant for analysis than the minimum energy flux principle.

However, this increase in efficiency at lower levels – reduction of entropy production per unit of mass of the metabolic system – will result in a higher stability of the metabolic system expressing the relative function in the long term only if the useful energy 'spared' at lower levels by higher efficiency is successfully moved up in the hierarchy (Margalef, 1968) and invested in the emergence of new structures/functions. This is the phenomenon of emergence which can be perceived only at the higher levels (by changing scale of observation), and only when using new formal identities (a different set of attributes for describing the performance, proxy variables and their relations) and inferential systems.

This cooperation between the minimum entropy production principle and the maximum energy flux principle requires the ability to effectively use what has been saved on the lower levels, due to efficiency improvements based on design, in order to invest in adaptability to changes of the context at the higher level. That is, at the higher level, improvements in efficiency obtained inside the system are transformed into the expression of new and more complex behaviours at the larger scale.

A very familiar example of this contrasting trade-off of efficiency across local-scale and large-scale levels can be obtained by looking at the household level or, for that matter, at the level of the firm within the domestic economy. In general terms households tend to make economies in their routine activities. That is, families tend to save on their daily shopping by looking for special offers, 'saving coupons' and cheaper supermarkets. On the other hand, as soon as the spare income reaches a certain threshold level, these savings are used to buy a larger car or a fancy vacation. Also in this case, there is a transfer across hierarchical levels of the gains obtained with efficiency. What is saved on the purchase of goods used within the routine metabolism of the household is transformed into a larger ability to make those investments which are able to enhance social interactions.

Put another way, the two trends of maximization of energy flux for the whole system (detected when describing the process of interaction of the whole with its context) and minimization of entropy production per unit of lower-level component (detected when describing the process of metabolism at a lower hierarchical level) are not exclusive of each other; rather they are operating in parallel on different scales *as far as the favourable boundary conditions are maintained.* The final outcome of these two trends is a better integration of metabolic systems with their environment during evolution. As suggested by Margalef and illustrated by the example of the household economy, by increasing the efficiency inside the system it is possible to match the continuous requirement for an increasing level of interaction with the context (for example changes in lifestyle). In this way internal characteristics and external characteristics of interacting metabolic systems are adjusted to each other in a process that should therefore be called co-evolution.

Another practical example can be obtained by studying the evolution in time of the characteristics of the energy metabolism of a given country by using simultaneously intensive and extensive variables and checking how the changes of the whole can be explained by changes in the profile of metabolism of the parts (the various economic sectors and sub-sectors).[24] Such an analysis of the energy metabolism of the Japanese economy between 1971 and 2001 is given in Figure 3.13. The variables included in the graph are:

- energy consumption per capita (EC_{PC}) – GJ/year per capita (an intensive variable);
- total energy consumption of the country (TEC) – EJ/year (an extensive variable);
- population – millions of Japanese (an extensive variable); and
- economic energy intensity (EEI) – this is the ratio MJ/yen (how much energy goes into the economy per one yen equivalent of goods and services) obtained by dividing TEC by GDP (an intensive variable).

All variables are normalized in terms of 1971 values = 100 per cent. Looking at the graph one can see after the first and second oil shocks – determining a situation of shortage in the supply of fossil energy – that technological improvements aimed at a more efficient use of energy resulted in a decrease in the value of EC_{PC} and TEC. Between 1971 and 1987, the effect of the two contrasting tendencies (saving

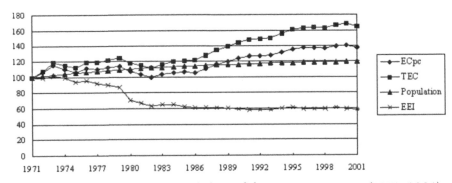

Figure 3.13   *The energy metabolism of the Japanese economy (1971–2001)*
in terms of four variables

Source: compiled from the data in Statistics Bureau (Ministry of Internal Affairs and Communications
of Japan) (2007)

energy in some activities in order to be able to allocate more energy in other
activities) can be detected by looking at the fluctuations (decreases and increases)
in the values of EC_{PC} and TEC. However, since 1988, both EC_{PC} and TEC have
been steadily increasing (except for a few years) more rapidly than the population
increase. Since 1986 the EEI (reflecting the efficiency of converting energy into
added value) has remained stable around 60 per cent. Therefore, according to this
variable it is not clear whether or not the increases in EC_{PC} and TEC since 1988
can be attributed directly to technological improvement per se. In any case, the
general trend of an overall increase in energy consumption per capita is clear.

The evolution in time of the energetic metabolism of Japan can also be
examined by looking at changes which took place at a level lower than the whole
country. This requires moving the analysis to the level of individual economic
sectors. As a matter of fact, it is possible to study the changes in the characteristics
of individual sectors using a combination of extensive variables – for example hours
of working time allocated, energy consumption and added value generated – and
intensive variables – for example energetic metabolic rate (energy consumed per
hour of labour), economic productivity per hour of labour (added value generated
per hour of labour) and economic energy intensity (energy consumed per yen of
added value generated). Therefore, when studying the evolution in time of the
metabolism of a socioeconomic system, we can also study changes in the
benchmark value characteristics of the various sectors. However, in order to relate
the changes taking place in individual sectors to the changes taking place at the
level of the whole economy, it becomes crucial to consider the relative importance
of the various sectors in the economic process. For example, in Figure 3.14 we can
see that different sectors of the Japanese economy went through important
changes in their economic energy intensity. However, in itself this information is
not enough to make an inference about the energy intensity of the whole

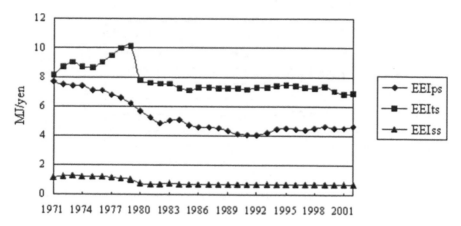

Figure 3.14   *Economic energy intensity of three sectors of the Japanese economy:*
PS (productive sector); TS (transport sector); SS (service sector)

Source: compiled from the data in Statistics Bureau (Ministry of Internal Affairs and Communications of Japan) (2007)

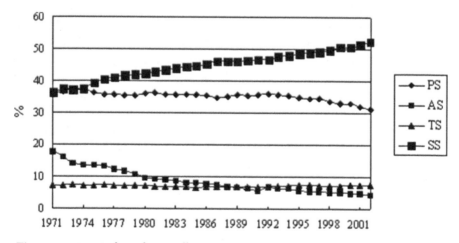

Figure 3.15   *Labour hours allocation to four sectors of the Japanese economy: PS*
(productive sector); AS (agricultural sector); TS (transport sector); SS (service sector)

Source: compiled from the data in Statistics Bureau (Ministry of Internal Affairs and Communications of Japan) (2007)

economy, unless we do not take into account the changes in the relative size of the various economic sectors – for example the distribution of economic activity and the workforce over the different sectors. In relation to the profile of the distribution of the work force, in Figure 3.15 we can see that Japan over this period of time followed a classic trend in the evolution of developed countries. The service sector keeps expanding in time, while the productive sectors, especially the agricultural sector (AS), are shrinking. From this perspective we can see that the Jevons Paradox is generated by a continuous change in the profile of allocation of available resources (human activity, capital and useful energy) over different economic sectors with different characteristics. The characteristics of the whole country, the overall process of becoming, can only be understood by looking at both the structural changes within various sectors (intensive variables defined at lower hierarchical levels) and the changes in the profile of distribution of the overall size of the whole economic process (characterized using extensive variables such as total human time, total capital and total energy consumption) over these lower-level elements (Giampietro and Mayumi, 2000a and b).

When interpreting sustainability within the concept of co-evolution (among different socioeconomic systems and ecological systems), the yin–yang tension entailed by the Jevons Paradox can be interpreted as dialectic tensions aimed at increasing the compatibility among the various systems of control operating on different space-time scales in either social or ecological systems. These systems of control are operating on different scales both within the system under analysis (for example a given economy) and within its environment, determining the favourable boundary conditions upon which that society relies for its survival (for example other economies and the various ecosystems embedding them). The need to balance these two requirements leads to a balance between the priorities given to these two principles.

Unfortunately, from an environmental point of view, in the past 200 years or so, concerning the growth of economic systems, we notice a tendency to give an excessive priority to the short-term increase in EFT2: greater speed of throughput in terms of production and consumption. Georgescu-Roegen describes human beings' addiction to the exosomatic comfort given by energy and mineral resources: 'exosomatism has also made us thoroughly addicted to the exosomatic comfort – hence almost completely dependent on the finite mineral dowry of our abode' (Georgescu-Roegen, 1992, p147). In the narrative of neoclassical economics, this tendency is compatible with a clear priority given to the acceleration of economic growth, and increasing GDP naturally becomes the main concern. In this perspective, the opposite concern with EFT1 at the large scale – the overall load of the ship in the metaphor proposed by Daly – is often neglected. This unbalanced tendency results in a low priority given to the goal of reducing the rate of depletion of natural resources and the levels of environmental pollution. The final result of the unbalance between these two principles is policy decisions based on the myopic rule 'the faster the throughput (GDP), the better'.

The yin–yang tension between efficiency and adaptability: Implications for sustainability

The discussion on the relationship between the minimum entropy production (or an interpretation of efficiency of type EFT1) and the maximum energy flux (or an interpretation of efficiency of type EFT2) can be easily connected with the two notions adopted in evolutionary views in ecological systems. Adaptability is a crucial quality for the sustainability of the evolution of ecological systems in the long term (Conrad, 1983; Ulanowicz, 1986; Holling, 1995). Here we adopt the following narrative for the definition of adaptability: the ability to adjust our own identity in order to retain fitness in the face of changing goals and changing constraints. Fitness means the ability to maintain congruence among a set of goals, the set of processes required to achieve them and constraints imposed by boundary conditions. Since metabolic systems are historically dependent, they preserve their individuality only if they manage to remain alive in the process of becoming. Therefore adaptability requires the ability to preserve diversity (an adequate option space) in terms of both possible behaviours and organizational structures.

However, the goal of preserving diversity per se collides with that of augmenting 'efficiency' as defined at a particular point in space and time, according to a particular interpretation of this term. That is, after having formalized what we mean by 'more efficiency' when performing a given function, we can finally rank different structural types by mapping onto the same function. At that point, in order to increase the efficiency of the process according to the chosen formalization, we have to eliminate those activities that are 'less effectively performing' and amplify those activities which are perceived as more effectively performing – for example the ballpoint pen replacing the pencil. This general rule drives technological progress. For example, in relation to the technological progress of agriculture, in the last century world agriculture was 'improved' according to the given set of goals expressed by those social groups in power and according to the given perception of boundary conditions (for example plenty of oil). As a consequence, the widespread adoption of the paradigm of industrial agriculture around the world brought about a dramatic reduction in the diversity of systems of production (for example the disappearance of many traditional farming systems). More effective techniques of agricultural production (following the interpretation of EFT2) were typified by monocultures of high-yielding varieties supported by 'energy intensive' technical inputs, such as synthetic fertilizers, pesticides and irrigation (Pimentel and Pimentel, 1996). Driven by technological innovations tailored on this interpretation of efficiency – such as the green revolution or genetically modified organisms – agricultural production all over our planet is converging on a very small set of standard solutions (commercial seeds, technological packages, and economic demand heavily affected by transnational corporations and globalized markets). On the other hand, the 'obsolete' agricultural systems of production, those that are being abandoned all over the planet, may show a very high performance if a different set of goals and criteria – rural employment, ecological compatibility and preservation of biodiversity – were adopted (Altieri, 1987). A similar 'return to the past' forced by a sudden change in boundary conditions was experienced by both NASA and the Soviet Space Agency, which had to take a step backwards in technology from the use of ballpoint pens to the use of pencils when the first astronauts started to operate in the absence of gravity.

When examining the process of evolution in terms of complex systems theory (see, for example, Kampis, 1991; Giampietro, 1997; Giampietro et al, 1997), we can observe that, in the last analysis, the drive toward instability is generated by the reciprocal influence between efficiency and adaptability, which are operating on different space-time scales. The continuous transformation of efficiency into adaptability and that of adaptability into efficiency is responsible for the continuous shift of the system into evolutionary trajectories as far as the boundary conditions are favourable. When choosing an arbitrary starting point, an impredicative loop of three steps is generated in the following way:

1   Accumulation of experience in social systems leads to more efficient processes of metabolism by amplifying more efficiently performing types and eliminating more poorly performing types in relation to a given definition of goals.

2   More efficient processes of metabolism make more surplus available to expand socioeconomic activities, implying the definition of new goals.

3   The consequent increase in the number of activities performed in a society and the scale of interaction of the socioeconomic system with its environment translate into an increased stress on the stability of boundary conditions, in other words more stress on the environment and a higher pressure on resources. This calls for increased investments in adaptability – for example the addition of other tasks for handling resource depletion and environmental pollution. However, in order to be able to invest more in adaptability the system needs to be more efficient. It has to better use its experience on how to carry out the required portfolio of tasks. Only when the investments in adaptability become successful can the system get back to step 1.

An overview of the coexistence of different possible causal paths between efficiency and adaptability, described on different timescales, is illustrated in Figure 3.16. When considered on a short timescale, efficiency would imply a negative effect on adaptability and vice versa. When a long-term perspective is adopted, they both thrive on each other. However, the only way to obtain a successful result based on a sound yin–yang tension is by continuously expanding the number of tasks performed by the system, in order to be able to expand the size of the domain of activity of human societies. That is, increases in efficiency are obtained by amplifying the more efficiently performing activities, without eliminating completely the obsolete ones. These activities will be preserved in the repertoire of possible activities of the societal system as a memory of 'different meanings' of efficiency. These different meanings could become useful again when facing a different set of boundary conditions or setting a different set of goals. Either new boundary conditions or new goals will require the execution of

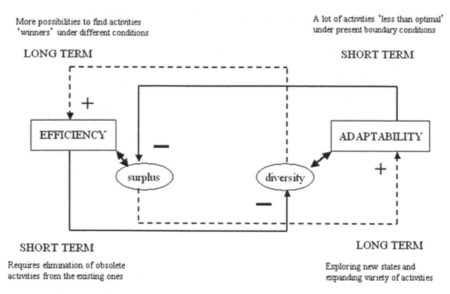

More possibilities to find activities
'winners' under different conditions

A lot of activities 'less than optimal'
under present boundary conditions

LONG TERM

SHORT TERM

EFFICIENCY

+

surplus

—

diversity

ADAPTABILITY

+

SHORT TERM

LONG TERM

Requires elimination of obsolete
activities from the existing ones

Exploring new states and
expanding variety of activities

Figure 3.16   *Self-entailment of efficiency and adaptability across scales*

new tasks, which in turn will require the adoption of different definitions of efficiency. After changing the definition of efficiency, the types used as a template for amplification until that moment will be considered as obsolete and will no longer be amplified. According to this new definition of efficiency, the system can look for new and more efficient types. In this situation, obsolete types still present in the available repertoire can be recycled and readopted in different situations or in relation to new goals. In this way, at each cycle, societal systems will enlarge their repertoire of knowledge of possible tasks to be carried out and of possible states to be accessed in this way. This expansion in the domain of activity of a society will translate into an increase in its size, no matter how we decide to measure its size – total amount of energy controlled, information processed, human mass or GNP (Giampietro et al, 1997).

Therefore the sustainability of societal systems can only be imagined as a dynamic balance between the rate of development of their efficiency and adaptability. A continuous update of the repertoire of structural types is required to maintain the existing repertoire of functions and a continuous update of the repertoire of functions is required to preserve the realization of existing structures. Put another way, neither a particular societal structure nor a particular societal function can be expected to be indefinitely sustainable in the future.

Practical solutions to this challenge require continuously deciding how to deal with 'the tragedy of change', an expression suggested by Funtowicz and Ravetz (1990), to deal with the predicament of sustainability. Any social system in its process of evolution has to decide how to become a different system, while maintaining its own individuality in this process. The 'feasibility' of this process – changing the structure of an airplane while flying on it – depends on the nature of internal and external constraints facing the society. The 'advisability' of the final changes – what the plane will look like at the end of the process, if still flying – will depend on the legitimate contrasting perceptions of those flying on it, their social and power relations, and the ability expressed by such a society to make wise changes to the plane at the required speed.

The tragedy of change represents an additional complication related to the process of decision making for sustainability. Namely, it is difficult to find an agreement on the set of the most important features to preserve or to enhance when attempting to build a different flying airplane. This has to do with how to define efficiency now. But this decision has to be taken without having reliable information about the feasibility of the various possible projects to be followed. As noted earlier, the definition and forecasting of viability constraints is unavoidably affected by a large dose of uncertainty and ignorance about the possible unexpected future situations. Put another way, when facing the sustainability predicament, humans must continuously gamble to try to find a balance in their definitions between efficiency and adaptability. In cultural terms, this means finding a point of equilibrium between the importance to be given to the past and the future when shaping the identity of their civilization (Giampietro, 1994).

CONCLUSION: THE IMPLICATIONS OF THE JEVONS PARADOX FOR SUSTAINABILITY ISSUES

So what? Is an increase in efficiency good or bad?

An improvement of efficiency in the set of technological processes sustaining society (for example more efficient cars) can generate two different results:

1   Benign for humans – when adopting a perception of improvement referring to the inside of the black box, an improvement of efficiency may be used to provide a better material standard of living for humans. In biophysical terms this efficiency improvement means having access to more energy and materials to be used in producing and consuming goods and services; and

2   Benign for ecological systems embedding the socioeconomic system – when adopting a perception of improvement referring to the environment of the black box, an improvement in efficiency may be used to reduce the level of natural resource consumption and the level of environmental impact. In biophysical terms this efficiency improvement means reducing the impact on the environment, in other words the environmental loading, associated with the extraction of resources and dumping of wastes from the economic process.

These two possible outcomes of technological improvement are described in Figure 3.17 in terms of movements on an EL–MSL plane. The horizontal axis refers to a proxy for environmental loading (EL). The vertical axis refers to a proxy for material standard of living (MSL). Depending on the particular issue considered we can have either a very simple or a more complicated formalization of the effects of energy efficiency. In relation to the mileage of a car, for instance, key attributes of performance (speed, payload, comfort) can be associated with MSL, and the gasoline requirement (fuel consumed per mileage or CO_2 emission per mileage) with EL. In the same way, when dealing with economic growth, we can imagine using an indicator of total energy consumption in the economy – highly correlated with GDP – as a proxy of MSL, and the total emissions of CO_2 as an indicator of EL. In more general terms, using the concepts previously discussed, we can say that increases in EFT2 or the rate of energy flux can be associated with upward movements along the MSL axis, while increases in EFT1 or the rate of entropy production can be associated with the leftward movements along the EL axis. Looking at the graph in Figure 3.17, and assuming point 1 as the initial position of the system, an increase in technological efficiency gives an increase in the degree of freedom for the system. That is, such an improvement provides the option of moving either from point 1 to point 1′ – associated with

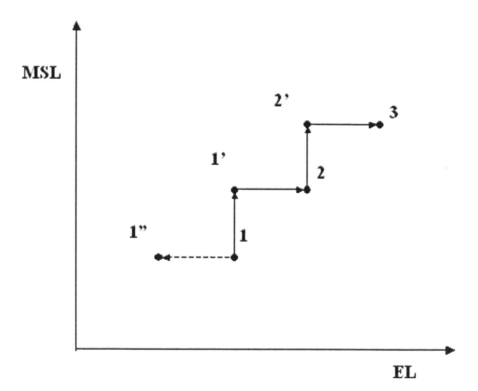

Figure 3.17   *The Jevons Paradox on an EL–MSL plane*

better services to humans (a higher value of MSL), while keeping the demand on the environment constant (the same value of EL) – or from point 1 to point 1″ – associated with the same level of services for humans (the same value of MSL), while reducing the demand from the environment (a lower value of EL). It is therefore the weight adopted in the political process of decision making that will decide whether technological improvement will be used to improve the welfare of the society or to reduce the pressure on the environment (Giampietro, 1994).

This point introduces a new element/dimension to our discussion of the Jevons Paradox, which is related to the existence of a systemic problem with the decision about how to take advantage of increases in efficiency. This systemic problem is generated by the existence of different hierarchical levels at which governance and policy implementation are evaluated and carried out in a given society. For example, let us imagine that a more efficient electric bulb (for example the bulb shown on the book cover) will enable the households of a given society to get a lower energy bill. As result of this change, families will have spare money for doing something else. The energy saved on the light bulb will be spent by the families on other types of activities. Again, at the level of society the gains obtained with efficiency will be moved to another sector and used to expand the option space of consumption. For this reason, if the priority of environmental protection is taken seriously, the government should either cap overall energy supply or introduce a tax proportional to the savings and *set aside this money*, in order to prevent the consumers from using the money saved by more efficient bulbs to fuel other activities. Otherwise, the saved money will be recycled into societal spending either by being used by the family or by the government taxing it. This will be analogous to the example of the more efficient engine driving the evolution of cars with a wider set of attributes of performance. An increase in efficiency makes it possible to do 'more of the same' in the short term and then a diversification in the pattern of final consumption in the long term. The policy relevance of this discussion is that, in order to be able to use a technical improvement in efficiency to get from point 1 to point 1″, national governments should tax more 'energy-efficient' appliances and use such a revenue only for expanding natural reserves – for example set aside forests in order to prevent more human development in the long term.

However, the evaluation of this choice will be different at different hierarchical levels – for example the level of the national government dealing with environmental impacts versus the level of the household dealing with their daily material standard of living. And although members of the household (and firms) are citizens as well as consumers (and producers), the majority of voters today would find it difficult to accept a solution in which increases in efficiency are used to tax households more and not to improve social services. This attitude of course reflects the status quo of voters' perceptions which, we hope, will change in the future, with an increase in environmental awareness and a better understanding of the interrelations among different types of decision-making processes at different levels. That is, we hope that in the future voters will elect governments using technical progress to limit their material standard of living for the sake of the environment. This is a crucial prerequisite to have governments that will be serious about the implementation of policies aimed at preserving the environment, rather than just paying lip service to it.

To date, however, technical progress has been perceived by humans, as a factor to be used to provide improvements for them and not for ecosystems. This is the reason why, up to now, technical progress followed the path $1 \rightarrow 1' \rightarrow 2 \rightarrow 2' \rightarrow 3$, with alternate increases in intensive and extensive variables driven by internal pressure. Gains have been transferred to humans rather than to natural systems. Historically, during their evolutionary process, human societies have steadily been moving in a northeastern direction in the MSL–EL plane. When extensive variables are also included in the picture, evolution tends to follow the trajectory $1 \rightarrow 2 \rightarrow 3$, as indicated in the three-dimensional space MSL–EL–SIZE illustrated in Figure 3.18. The term SIZE here can be interpreted in biophysical terms as a larger amount of either energy flow, matter flow or space utilization by the socioeconomic system in its process of autopoiesis. When considering this three–dimensional representation, the trajectory is not regular and cannot be predicted in relation to the relative weight of changes in intensive and extensive variables. What matters, however, in relation to our discussion of the Jevons Paradox is that this trajectory, so far, keeps moving socioeconomic systems toward higher values of MSL, EL and SIZE in consequence of the dramatic technological progress of the last two centuries or so. Again the technological fix – including the ability to improve

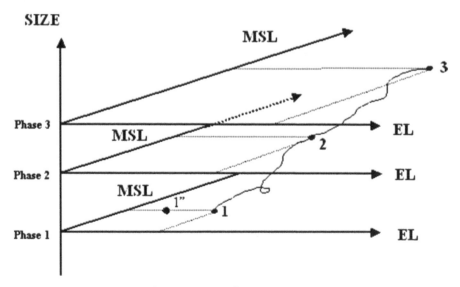

Figure 3.18   *The Jevons Paradox in an EL–MSL–SIZE space*

efficiency – does not, per se, automatically bring about a solution to the issue of sustainability of human development.

So what? How can quantitative analysis be used in sustainability science?

Any given perception/representation of the external world based on a particular formal model must necessarily reflect a set of choices made by a special story-teller about the selection of a relevant narrative for a given state of affairs in relation to a given set of goals (Rosen, 2000; Giampietro, 2003; Mayumi and Giampietro, 2006). As Schumpeter aptly remarked, '[a]nalytical work begins with material provided by our vision of things, and this vision is ideological almost by definition' (Schumpeter, 1954, p42).

Therefore, all models and indicators (data) do reflect a pre-analytical choice on *how to represent in a finite and closed information space* the external world and not the external world itself. This entails that complex events occurring across multiple scales cannot be fully perceived and represented by using a formal language (Giampietro et al, 2006a). Therefore it is absolutely necessary to perform a dual quality check: not only does the selected model have to be pertinent in relation to the chosen narrative, but the selected narrative must also carry relevance for those using the results of the analysis. In this perspective Box's maxim should be recalled: 'all models are wrong, but some are useful' (Box, 1979). That is, a quality check on the usefulness of models has to do first of all with their relevance. A model based on an irrelevant narrative is much worse than a model based on a wrong formalism. Such a model will keep providing irrelevant and misleading conclusions for ever.

Funtowicz and Ravetz (1990) developed a new epistemological framework called post-normal science (PNS). In PNS, uncertainty, the story-telling associated with different stakeholders and their value conflicts should be considered as crucial elements in the process of decision making. The adjective 'post-normal' indicates a departure from curiosity-driven or puzzle-solving exercises of normal science in the Kuhnian sense (Kuhn, 1962). Normal science was successfully extended from the laboratory of core science to the conquest of nature through applied science. However, this 'normal' scientific approach is no longer appropriate for the solution of sustainability problems. In fact, the social, technical and ecological dimensions of sustainability problems are so deeply connected that it is simply impossible to consider these dimensions separated, one at the time, as conducted in the analyses and assessments of conventional disciplinary fields.

However, in order to be able to find a solution for a problem, it is crucial to start with the acknowledgement that the problem exists in the first place. This is the reason why in relation to the Jevons Paradox, we have provided in this chapter a long epistemological discussion trying to make the point that conventional analytical tools based on the strategy of simplification typical of reductionism cannot be applied to the quantitative analysis of evolution necessary for sustainability issues. On the other hand, one must be aware of the fact that biophysical constraints interpreted using physical laws will always shape our option space. No matter what we would like to get from our new flying plane, it must match the laws of aerodynamics if it is to fly.

Rather than using models and numbers to predict the future and to find 'optimal solutions' and 'best courses of action', the proposed approach uses models and numbers to check the quality of the narratives selected in the pre-analytical step, which are used to characterize a given situation or to define possible scenarios. This paradigm shift follows the suggestion of Simon that when dealing with governance and sustainability science people should move away from substantive rationality and look rather for procedural rationality (Simon, 1976 and 1996). Such a change of paradigm obviously requires a drastic change in both the procedure and the analytical tool-kit adopted for the analysis.

The Jevons Paradox: Practical lessons for the analyst

To conclude this theoretical chapter we provide five simple and heuristic rules to be followed by scientists willing to apply quantitative analysis to complex adaptive systems in relation to sustainability issues:

1   Avoid formalism nonsense. Before starting 'crunching numbers' based on any chosen formalization, it is important to examine whether or not the underlying narratives behind the formalism are relevant for those that will use the results of the model. Numbers without a robust external referent are just human-generated constructs, which can generate misleading conclusions. It is not syntactic rigour of the formalism per se, but the validity of the particular narrative associated with issue definition and problem structuring, as well as the soundness of the measurement scheme that make numbers meaningful.
2   Look for integrated analysis across dimensions and scales. A useful mix of relevant types and categories, belonging to non-equivalent descriptive domains, is required to characterize the object of study in relation to the different narratives and scales relevant for sustainability analysis. After checking that selected types and categories are relevant and pertinent, it is then necessary to check the congruence of the resulting non-equivalent representations across scales and dimensions.
3   Always remember to put the observer's role back into the picture. When dealing with sustainability issues, there are three entities evolving together:
    a   the society embedding the modeller (the process choosing the modellers);
    b   the modeller studying a system in question (the process choosing the models); and
    c   the modelled system (the observed process becoming something else).

Sustainability analysis requires the ability to tailor in an adequate time both the issue definition and the problem structuring used to perform an integrated assessment on what is relevant, credible and acceptable for the social actors in a given context in relation to the three entities above.

4   Use numbers not to individuate 'the best course of action' but to check the quality of the underlying narratives. Sustainability analysis requires mediating between non-equivalent perspectives (associated with different story-tellers), which requires the use of different narratives about it. Therefore quantitative analyses can be used to check the quality of narratives, but not to look for optimal courses of action. Optimization protocols per se do not indicate the best course of action. They only reflect a ranking determined by the narrative selected for the quantification.

5   Remember that any system under investigation is special, meaning that nobody can predict its future. To explain mechanisms of causality, existing trends and equivalence classes we can use expected relations over types. This makes it possible for us to find, in time, useful explanations, predictive models and statistical regularities. However, perception and representation are necessarily based on types. Therefore they reflect only partially the external world, which is made of individual instances of these types. Sustainability has to do with things that are happening for the first time and only once. Georgescu-Roegen's observation should be recalled here: 'The predicament of the evolutionary biologists is that he has never observed another human species being born, aging and dying' (1975a, p349). Life is the interaction of autopoietic systems which base the definition of their identity on impredicativity. For this reason nobody can predict their evolution.

ACKNOWLEDGEMENT

Mario Giampietro gratefully acknowledges the financial support of the European Commission 6th Framework programme for the activities of the European Project DECOIN – FP6 2005-SSP-5-A: 044428. We are grateful to the US Department of Energy and Eartheasy.com for allowing us to reproduce the two diagrams in Figure 3.12.

NOTES

1 According to the definition given by Wikipedia: 'Complex adaptive systems are special cases of complex systems. They are *complex* in that they are diverse and made up of multiple interconnected elements and *adaptive* in that they have the capacity to change and learn from experience. The term *complex adaptive systems* was coined at the interdisciplinary Santa Fe Institute (SFI), by John H. Holland, Murray Gell-Mann and others.' In our interpretation the class of complex adaptive systems should be

characterized by using four key characteristics: (1) they are open systems which cannot be in thermodynamic equilibrium; (2) they are hierarchically organized and operating on multiple spatial-temporal scales; (3) they are autopoietic systems (their definition entails impredicativity); and (4) they are organized in a particular type of nested hierarchy which can be perceived and represented only using the concept of 'holon'. These concepts are discussed later in the chapter.

2 *Extensive variables* are additive variables which are used to quantify the size of a system in relation to a relevant observable quality (for example kilograms of mass or litres of volume); *intensive variables* are non-additive variables which are used to quantify a relevant quality of a system per unit of its size (for example degrees C of temperature or millibars of pressure).

3 Here we mean by reductionism the general tenet in science and philosophy that the nature of complex things can be reduced to the nature of sums of simpler or more fundamental things, whose description can be explained in terms of linear causations. This tenet can be said of objects, phenomena, explanations, theories and meanings. However, we should remember that Aristotle strongly opposed this tenet in his *Metaphysics*: 'In the case of all things that have several parts and in which the whole is not like a heap, but is a particular something besides the parts, there is some such uniting factor' (Book Eta). The great theoretical physicist Max Planck shares the same view with Aristotle: 'the whole is never equal simply to the sum of its various parts' (Planck, 1959, p255).

4 For our present purpose we may be satisfied with the simple definition of entropy as an index of the amount of unavailable energy in a given system at a given moment (Mayumi, 2005, p148). For a precise thermodynamic definition, see Fermi (1936, Chapter IV). For technical discussions of the entropy concept applied to the analysis of sustainability, see Mayumi (2001), Giampietro and Mayumi (2004), and Mayumi and Giampietro (2004).

5 Perhaps the stable trend of EC_{PC} for the US economy might be explained by the increase in the population of lower income families, mainly by immigration, and its effect on the skewed income distribution within the US economy.

6 Georgescu-Roegen states that the 'institutions of the market, money, credit and enterprises of all sorts emerged in response to the progressive evolution of the exosomatic nature of *homo sapiens sapiens*' (1986). However, his idea of bioeconomics is different from those of some people who tried to reduce every economic phenomenon into biological analogies: 'my use of "bioeconomics" had not been influenced by the prevailing intellectual fashion of reducing everything to a biological basis' (Georgescu-Roegen, 1986).

7 Takeshi Murota calls the Jevons Paradox the 'Jevons law on economy of fuel' in an interesting paper on gas hydrate exploration and its environmental consequences (Murota, 1996).

8 There is difficulty in ascertaining what Say's Law of Markets really means – for readers interested in this issue, see Baumol (1977). Here we follow Schumpeter's interpretation:

> ... [social totals of] aggregate demand and aggregate supply are not independent
> of each other, because the component demands 'for the output of any industry
> (or firm or individual) comes from the supplies of all the other industries

(or firm or individual)' and therefore will in most cases increase (in real terms)
if these supplies increase and decrease if these supplies decrease. (Schumpeter,
1954, p617)

9 An interesting definition of emergence is provided by Jeffrey Goldstein in the
inaugural issue of *Emergence*. According to him, emergence 'refers to the arising of
novel and coherent structures, patterns and properties during the process of self-
organization in complex systems. Although emergent phenomena appear differently
in different types of systems [for example whether they occur in physical systems or
in computer simulations] they share certain interrelated, common properties that
identify them as emergent' (Goldstein, 1999, p49). However, our standpoint is a bit
different from Goldstein's, since he states that 'emergence is emerging today as a
construct of complex, dynamical systems'. In our view, no dynamical system can
exhibit the phenomena of emergence discussed in this chapter (Georgescu-Roegen,
1975b; Giampietro et al, 2005).

10 *Catch-22* is the title of Joseph Heller's well-known anti-war novel where a paradox of
military regulations is provided (Heller, 1961). There is an easy way for a terrified
bomber pilot to get out of duty. He can ask for a psychological discharge on the
grounds that he is crazy. However, nobody but an insane person would actually want
to fly the suicidal attacks the squadron is assigned. Thus, anyone who asks to get out
of duty is, by definition, not crazy, so he must fly and get back into the war!

11 A thermodynamic state is the ensemble of all the dynamical states through which, as
a result of molecular motion, the system is rapidly passing. The states of equilibrium
have the property of not varying so long as the external conditions remain unchanged.
Thus the gas enclosed in a container of constant volume is in equilibrium when its
pressure is constant throughout and its temperature is equal to that of the
environment (Fermi, 1936). Prigogine and his co-workers have tried to extend the
methods of thermodynamics to treat the entire range of phenomena starting from
equilibrium and including non-linear situations and instabilities (Prigogine, 1961;
Glansdorff and Prigogine, 1971; Nicolis and Prigogine, 1977). However, their
analyses are basically restricted to the class of situations for which the local entropy
may be represented in terms of the same independent variables *as if the system were at
equilibrium.* This type of restriction is too severe to deal with real living systems.

12 The term 'substantive' refers to something that exists in its own right. In relation to
the term, Herbert Simon (1976) distinguishes substantive rationality in conventional
economics from procedural rationality within psychology. The former rationality
refers to behaviour 'when it is appropriate to the achievement of given goals within
the limits imposed by given conditions and constraints'. The latter rationality refers
to behaviour 'when it is the outcome of appropriate deliberation' and depends on the
process that generates the behaviour.

13 Georgescu-Roegen (1971) introduced the term 'arithmomorphic' based on the
properties of the real number system: within the continuum every real number retains
a distinct individuality. So arithmomorphic concepts do not overlap. It is this very
restrictive property of the real number system with which mathematical logic works
with tremendous efficiency. However, because of this peculiar nature, the
arithmormorphic models cannot deal with real qualitative changes in evolving
metabolic systems.

14 Russell's notion of vicious circle comes from his discovery of a paradox in set theory. A set can easily have elements that are themselves sets, for example {1, {2, 3}, 5}. We call a set that contains itself as one of its elements an abnormal set, and any set that does not contain itself as an element we call a normal set. Most sets are normal, and if we suspect that abnormal sets are undesirable, we might try to confine our attention only to the set S of all normal sets. It is relatively easy to show that if S is normal, then it must be abnormal, and that if S is abnormal, then it must be normal! (Wilder, 1952) This is the vicious circle created by Russell's paradox.

15 A detailed discussion of the concept of holon in relation to standard epistemological predicaments associated with multiple scales and complexity is provided in Giampietro et al (2006a).

16 Salthe (1985) suggests two concepts: types and individual instances. A type is an equivalent class of entities (in terms of either structures or functions) that can get reproduced or realized within a stable associative context. An individual instance is a special realization of a particular type within a specific context. There is a difference between a type and an individual instance (realization or fabrication) of the type. The information associated with the essence of the type is an expected set of attributes known by the observer, whereas a given instance of the type is a natural system that only matches the expected set of attributes to a certain extent.

17 Simon (1962) casts this issue in terms of the ability of wisely coupling *organized structure* to *relational function*. Bailey (1990) proposes the same approach for social systems using the couplet of terms *incumbent* and *role*.

18 Rosen (2000) proposes, within a more general theory of modelling relation – adopted by autopoietic systems – a more drastic distinction which gets back to the Greek philosophical tradition. Rosen makes a distinction of this dual description based on: 'individual realizations' and *essences*. In this distinction, realizations are always local and 'special'. They cannot be captured and fully described by any scientific representation, because any individual realization maps only imperfectly onto its relative type. That imperfection comes from the unique history of each realization. On the other hand, the essence – generated by the ability of sharing a commensurate experience within a given knowledge system – is associated with the expected characteristics of an equivalence class of realizations mapping onto a type which has been judged relevant for a knowledge system.

19 Tagore insists on the importance of a relevant story-teller in his dialogue about science and realism with Einstein:

> *This world is a human world – the scientific view of it is also that of scientific man. Therefore the world apart from us does not exist. It is a relative world, depending for its reality upon our consciousness. … The table is that which is perceptible by some kind of consciousness we possess. … If there be any truths absolutely unrelated to humanity, then for us it is absolutely non-existing.* (Home and Robinson, 1995, pp174–175)

20 The 'Carnot cycle' is the archetypical reversible cycle, and a Carnot engine is one that does not dissipate any energy internally and uses only reversible steps. A Carnot cycle consists of four steps: (1) isothermal expansion in contact with the hot reservoir; (2) adiabatic expansion after the hot reservoir is removed; (3) isothermal compression in

contact with the cold reservoir; and (4) adiabatic compression after the cold reservoir is removed. Carnot discovered that no real heat engine operating between a hot reservoir at temperature T_2 and a cold reservoir at temperature T_1 can be more efficient than a Carnot engine operating between those two reservoirs. The efficiency of a Carnot engine is $(T_2-T_1)/T_2$. Therefore this efficiency depends only on the two temperatures.

21 Prigogine himself acknowledged the severe limitations of his approach:

> ... the Gibbs formula was originally proved for the equilibrium conditions, and its use for the non-equilibrium conditions is a new postulate on which the whole of the thermodynamics of irreversible processes is based. The physical interpretation of this basic formula is that, even without equilibrium, the entropy depends only on the same independent variables as for equilibrium processes. This is certainly not true [for] very far from equilibrium. (Prigogine, 1961, p93)

22 When the high level of production is attained after the industrial revolution, keeping high levels of consumption became a major concern for economists. Georgescu-Roegen notes the following:

> In those 'land of plenty' the consumers became 'king', as the new situation has so often been characterized. For the interest of productive activities in general it seemed that what was scarce was the demand for each kind of product. This is how utility came to be regarded as the source of value. Neoclassical economics, the new doctrine moulded on an economic reality of abundance, began teaching that the structure of the economic process is determined in mechanistic way by the relative importance people attribute to the enjoyments of different commodities and the drudgeries of various kind of work. (Georgescu-Roegen, 1982, pp3–4)

23 A similar view is given by Maturana and Varela:

> ... any given organization may be realized through many different structures, and that different subsets of relations included in the structure of a given entity may be abstracted by an observer (or its operational equivalent) as organizations that define different classes of composite unities. (Maturana and Varela, 1980, pxx).

24 In relation to this task Giampietro and Mayumi proposed a methodology called multi-scale integrated analysis of societal metabolism (MSIASM) (Giampietro and Mayumi, 1997, 2000a, b; see also Giampietro, 2003; Giampietro et al, 2006b; Ramos-Martin et al 2007). This methodology is based on pioneering work by Zipf (1941), Lotka (1956), Georgescu-Roegen (1971 and 1976) and Ulanowicz (1986). One of the theoretical pillars of MSIASM is that the technological development of a society can be described in terms of an acceleration of energy and material consumption in the productive sectors, together with the dramatic reallocation of distribution of age classes, human time profile of activities and land-use patterns among the different sectors of the modern economy. The overall result is a dramatic reduction of the working time allocated to the energy and agricultural sectors.

REFERENCES

Ahl, V. and Allen, T. F. H. (1996) *Hierarchy Theory: A Vision, Vocabulary and Epistemology*, Columbia University Press, New York

Alcott, B. (2005) 'Jevons' paradox', *Ecological Economics*, 54, pp9–21

Allen, T. F. H. and Starr, T. B. (1982) *Hierarchy*, University of Chicago Press, Chicago, IL

Altieri, M. (1987) *Agroecology: The Scientific Basis for Alternative Agriculture*, Westview Press, Boulder, CO

Annual Energy Review (2006), www.eia.doe.gov/emeu/aer/contents.html

Aristotle (edition 1960) *Metaphysics*, translated by R. Hope, University of Michigan Press, Ann Arbor, MI

Bailey, K. D. (1990) *Social Entropy*, State University of New York Press, Albany, NY

Baumol, W. J. (1977) 'Say's (at least) eight laws, or what Say and James Mill may really have meant', *Economica*, vol 44, pp145–162

Box, G. (1979) 'Robustness is the strategy of scientific model building', in R. L. Launer and G. N. Wilkinson (eds) *Robustness in Statistics*, Academic Press NY, New York, pp201–236

Brookes, L. A. (1979) '"A low-energy strategy for the UK" by G. Leach et al: A review and reply', *Atom*, vol 269, pp3–8

Brown, J. H. and West, G. B. (2000) *Scaling in Biology*, Oxford University Press, Oxford

Carpenter, S. R. and Kitchell, J. F. (1987) 'The temporal scale of variance in limnetic primary production', *American Naturalist*, vol 129, pp417–433

Cherfas, J. (1991) 'Skeptics and visionaries examine energy saving', *Science*, vol 251, pp154–156

Clark, B. and Foster, J. B. (2001) 'William Stanley Jevons and the coal question: An introduction to Jevons's *Of the Economy of Fuel*', *Organization & Environment*, vol 14, pp93–98

Conrad, M. (1983) *Adaptability: The Significance of Variability from Molecule to Ecosystem*, Prenum, New York

Daly, H. (1995) 'Consumption and welfare: Two views of value added', *Review of Social Economy*, vol 53, pp451–473

Daly, H. (1996) *Beyond Growth: The Economics of Sustainable Development*, Beacon Press, Boston, MA

FAOSTAT (2007), http://faostat.fao.org/ (Food Balance Sheet)

Fermi, E. (1936) *Thermodynamics*, Dover, New York

Funtowicz, S. O. and Ravetz, J. R. (1990) 'Post-normal science: A new science for new times', *Scientific European*, vol 266, pp20–22

Gell-Man, M. (1994) *The Quark and the Jaguar*, Freeman, New York

Georgescu-Roegen, N. (1971) *The Entropy Law and the Economic Process*, Harvard University Press, Cambridge, MA

Georgescu-Roegen, N. (1975a) 'Energy and economic myths', *Southern Economic Journal*, vol 41, pp347–381

Georgescu-Roegen, N. (1975b) 'Dynamic models and economic growth', *World Development*, vol 3, pp765–783

Georgescu-Roegen, N. (1976) *Energy and Economic Myths: Institutional and Analytical Essays*, Pergamon Press, New York

Georgescu-Roegen, N. (1982) 'The energetic theory of value: A topical economic fallacy', Working Paper No 82-W16, Department of Economics, Vanderbilt University, Nashville, TN

Georgescu-Roegen, N. (1986) 'Man and production', in M. Baranzini and R. Scazzieri (eds) *Foundations of Economics: Structures of Inquiry and Economic Theory*, Basil Blackwell, Oxford, UK

Georgescu-Roegen, N. (1992) 'Nicholas Georgescu-Roegen about himself', in M. Szenberg (ed) *The Life Philosophies of Eminent Economists*, Cambridge University Press, New York

Giampietro, M. (1994) 'Using hierarchy theory to explore the concept of sustainable development', *Futures*, vol 26, pp616–625

Giampietro, M. (1997) 'Linking technology, natural resources and the socioeconomic structure of human society: A theoretical model', in L. Freese (ed) *Advances in Human Ecology*, Volume 6, JAI Press, Greenwich, CT

Giampietro, M. (1998) 'Energy budget and demographic changes in socioeconomic systems', in S. Dwyer, U. Ganslosser and M. O'Connor (eds) *Life Science Dimensions: Ecological Economics and Sustainable Use*, Filander Verlag, Fürth, Germany

Giampietro, M. (2003) *Multi-Scale Integrated Analysis of Agro-ecosystems*, CRC Press, Boca Raton, FL

Giampietro, M. (2006) 'Comments on "The energetic metabolism of the European Union and the United States" by Haberl and colleagues: Theoretical and practical considerations on the meaning and usefulness of traditional energy analysis', *Journal of Industrial Ecology*, vol 10, pp173–185

Giampietro, M. and Mayumi, K. (1997) 'A dynamic model of socioeconomic systems based on hierarchy theory and its application to sustainability', *Structural Change and Economic Dynamics*, vol 8, pp453–469

Giampietro, M. and Mayumi, K. (1998) 'Another view of development, ecological degradation and North–South trade', *Review of Social Economy*, vol 56, pp20–36

Giampietro, M. and Mayumi, K. (2000a) 'Multiple-scale integrated assessment of societal metabolism: Introducing the approach', *Population and Environment*, vol 22, pp109–153

Giampietro, M. and Mayumi, K. (2000b) 'Multiple-scale integrated assessment of societal metabolism: Integrating biophysical and economic representations across scales', *Population and Environment*, vol 22, pp155–210

Giampietro, M. and Mayumi, K. (2004) 'Complex systems and energy', in C. Cleveland (ed) *Encyclopedia of Energy*, Volume 1, Elsevier, San Diego, CA

Giampietro, M., Bukkens, S. G. F. and Pimentel, D. (1997) 'Linking technology, natural resources, and the socioeconomic structure of human society: Examples and applications', in L. Freese (ed) *Advances in Human Ecology*, Volume 6, JAI Press, Greenwich, CT

Giampietro M., Mayumi K. and Pimentel D. (2005) 'Mathematical models of society and development: Dealing with the complexity of multiple-scales and the semiotic process associated with development', in Jerzy Filar (ed) *Mathematical Models of Society and Development [Theme 6.3]*, Encyclopedia Of Life Support Systems (EOLSS), developed under the auspices of the UNESCO, EOLSS Publishers, Oxford, UK, www.eolss.net

Giampietro, M., Allen, T. F. and Mayumi, K. (2006a) 'The epistemological predicament associated with purposive quantitative analysis', *Ecological Complexity*, vol 3, pp307–327

Giampietro, M., Mayumi, K. and Munda, G. (2006b) 'Integrated assessment and energy analysis: Quality assurance in multi-criteria analysis of sustainability', *Energy*, vol 31, pp59–86

Glansdorff, P. and Prigogine, I. (1971) *Thermodynamics Theory of Structure, Stability and Fluctuations*, John Wiley & Sons, New York

Goldstein, G. J. (1999) 'Emergence as a construct: History and issues', *Emergence*, vol 1, pp49–72

Heller, J. (1961) *Catch-22*, Simon and Schuster, New York

Herring, H. (1999) 'Does energy efficiency save energy? The debate and its consequences', *Applied Energy*, vol 63, pp209–226

Heston, A., Summers, R. and Aten, B. (2006) *Penn World Table Version 6.2*, Center for International Comparisons of Production, Income and Prices at the University of Pennsylvania, September

Holling, C. S. (1995) 'Biodiversity in the functioning of ecosystems: An ecological synthesis', in C. Perrings, K. G. Maeler, C. Folke, C. S. Holling and B. O. Jansson (eds) *Biodiversity Loss: Economic and Ecological Issues*, Cambridge University Press, Cambridge, UK

Home, D. and Robinson, A. (1995) 'Einstein and Tagore: Man, nature and mysticism', *Journal of Consciousness Studies*, vol 2, pp167–179

Jevons, F. (1990) 'Greenhouse – A paradox', *Search*, vol 21, pp171–172

Jevons, W. S. (1865) *The Coal Question* (reprint of the 3rd edition – 1906), Augustus M. Kelley, New York

Jørgensen, S. E. (1992) *Integration of Ecosystem Theories: A Pattern*, Kluwer Academic Publishers, Dordrecht, The Netherlands

Kampis, G. (1991) *Self-Modifying Systems in Biology and Cognitive Science: A New Framework for Dynamics, Information and Complexity*, Pergamon Press, New York

Kawamiya, N. (1983) *Entropii to Kougyoushakai no Sentaku* [*Entropy and Future Choices for the Industrial Society*], Kaimei, Tokyo (in Japanese)

Khazzoom, J. D. (1980) 'Economic implications of mandated efficiency standards for household appliances', *Energy Journal*, vol 1, pp21–39

Khazzoom, J. D. (1987) 'Energy saving resulting from the adoption of more efficient appliances', *Energy Journal*, vol 8, pp85–89

Kleene, S. C. (1952) *Introduction to Metamathematics*, D. Van Nostrand, London

Koestler, A. (1967) *The Ghost in the Machine*, MacMillan, New York

Koestler, A. (1969) 'Beyond atomism and holism – The concept of the holon', in A. Koestler and J. R. Smythies (eds) *Beyond Reductionism*, Hutchinson, London

Koestler, A. (1978) *Janus: A Summing Up*, Hutchinson, London

Kuhn, T. S. (1962) *The Structure of Scientific Revolutions*, The University of Chicago Press, Chicago, IL

Layzer, D. (1988) 'Growth of order in the universe', in B. H. Weber, D. J. Depew and J. D. Smith (eds) *Entropy, Information and Evolution*, MIT Press, Cambridge, MA

Lietz, P. and Streicher, T. (2002) 'Impredicativity entails untypedness', *Mathematical Structures in Computer Science*, vol 12, pp335–347

Lotka, A. J. (1922) 'Contribution to the energetics of evolution', *Proceedings of National Academy of Sciences*, vol 8, pp147–151

Lotka, A. J. (1956) *Elements of Mathematical Biology*, Dover, New York

Mandelbrot, B. B. (1967) 'How long is the coast of Britain? Statistical self-similarity and fractal dimensions', *Science*, vol 155, pp636–638

Margalef, R. (1968) *Perspectives in Ecological Theory*, The University of Chicago Press, Chicago, IL

Maturana, H. R. and Varela, F. J. (1980) *Autopoiesis and Cognition: The Realization of the Living*, D. Reidel Publishing, Dordrecht, The Netherlands

Maturana, H. R. and Varela, F. J. (1998) *The Tree of Knowledge: The Biological Roots of Human Understanding*, Shambhala Publications, Boston, MA

Mayumi, K. (2001) *The Origins of Ecological Economics: The Bioeconomics of Georgescu-Roegen*, Routledge, London

Mayumi, K. (2005) 'Entropy', in C. J. Cleveland and C. G. Morris (eds) *Dictionary of Energy*, Elsevier, Amsterdam

Mayumi, K. and Giampietro, M. (2004) 'Entropy in ecological economics', in J. Proops and P. Safonov (eds) *Modelling in Ecological Economics*, Edward Elgar, Cheltenham, UK

Mayumi, K. and Giampietro, M. (2006) 'The epistemological challenge of self-modifying systems: Governance and sustainability in the post-normal science era', *Ecological Economics*, vol 57, pp382–399

Mayumi, K., Giampietro, M. and Gowdy, J. M. (1998) 'Georgescu-Roegen/Daly versus Solow/Stiglitz revisited', *Ecological Economics*, vol 27, pp115–117

Morowitz, H. J. (1979) *Energy Flow in Biology*, Ox Bow Press, Woodbridge, CT

Murota, T. (1996) 'Gas hydrate exploration: Its technology–environment interface in the world and Japan', *Hitotsubashi Journal of Economics*, vol 37, pp21–44

Newman, P. (1991) 'Greenhouse, oil and cities', *Futures*, vol 5, pp335–348

Nicolis, G. and Prigogine, I. (1977) *Self-Organization in Nonequilibrium Systems*, John Wiley and Sons, New York

Odum, H. T. (1971) *Environment, Power and Society*, Wiley-Interscience, New York

Odum, H. T. (1996) *Environmental Accounting, Emergy and Decision-making*, John Wiley, New York

Odum, H. T and Pinkerton, R. C. (1955) 'Time's speed regulator: The optimum efficiency for maximum power output in physical and biological systems', *American Scientist*, vol 43, pp331–343

O'Neill, R. V. (1989) 'Perspectives in hierarchy and scale', in J. Roughgarden, R. M. May and S. Levin (eds) *Perspectives in Ecological Theory*, Princeton University Press, Princeton, NJ

O'Neill, R. V., DeAngelis, D. L., Waide, J. B. and Allen, T. F. H. (1986) *A Hierarchical Concept of Ecosystems*, Princeton University Press, Princeton, NJ

Peters, R. H. (1986) *The Ecological Implications of Body Size*, Cambridge University Press, Cambridge, UK

Pimentel, D. and Pimentel, M. (1996) *Food, Energy and Society*, University Press of Colorado, Niwot, CO

Planck, M. (1959) *The New Science*, Meridian Books, New York

Plenty-Coups (2002) *Plenty-Coups: Chief of the Crows*, Bison Books, Winnipeg, Canada

Polimeni, J. M. and Polimeni, R. I. (2006) 'Jevons' paradox and the myth of technological liberation', *Ecological Complexity*, vol 3, pp344–353

Prigogine, I. (1961) *Introduction to Thermodynamics of Irreversible Processes*, 2nd revised edition, Interscience Publishers, New York

Prigogine, I. (1987) *From Being to Becoming*, Freeman, San Francisco, CA

Ramos-Martin, J., Giampietro, M. and Mayumi, K. (2007) 'On China's exosomatic energy metabolism: An application of multi-scale integrated analysis of societal metabolism (MSIASM)', *Ecological Economics*, vol 63, pp174–191

Rosen, R. (1977) 'Complexity as a system property', *International Journal of General Systems*, vol 3, pp227–232

Rosen, R. (1985) *Anticipatory Systems: Philosophical, Mathematical and Methodological Foundations*, Pergamon Press, New York

Rosen, R. (2000) *Essays on Life Itself*, Columbia University Press, New York

Salthe, S. N. (1985) *Evolving Hierarchical Systems*, Columbia University Press, New York

Saunders, H. (2000) 'A view from the macro side: Rebound, backfire and Khazzoom-Brookes', *Energy Policy*, vol 28, pp439–449

Schneider, E. D. and Kay, J. J. (1994) 'Life as a manifestation of the second law of thermodynamics', *Mathematical and Computer Modelling*, vol 19, pp25–48

Schumpeter, J. A. (1954) *History of Economic Analysis*, George Allen & Unwin, London

Simon, H. A. (1962) 'The architecture of complexity', *Proceedings of the American Philosophical Society*, vol 106, pp467–482

Simon, H. A. (1976) 'From substantive to procedural rationality', in S. J. Latsis (ed) *Methods and Appraisal in Economics*, Cambridge University Press, Cambridge, UK

Simon, H. A. (1996) *The Sciences of the Artificial* (3rd edition), MIT Press, Cambridge, MA

Smil, V. (1994) *Energy in World History*, Westview Press, Boulder, CO

Statistics Bureau (Ministry of Internal Affairs and Communications of Japan) (2007) Director General for Policy Planning and Statistical Research and Training Institute, www.stat.go.jp/english/index.htm

Ulanowicz, R. E. (1986) *Growth and Development: Ecosystem Phenomenology*, Springer-Verlag, New York

Whyte, L. L., Wilson, A. G. and Wilson, D. (eds) (1969) *Hierarchical Structures*, Elsevier, New York

Wilder, R. L. (1952) *Introduction to the Foundations of Mathematics*, John Wiley and Sons, New York

Zipf, G. K. (1941) *National Unity and Disunity: The Nation as a Bio-social Organism*, Principia Press, Bloomington, IN

4

Empirical Evidence for the Jevons Paradox

INTRODUCTION

This chapter builds upon Chapter 2, which gives the historical perspective of the Jevons Paradox, and Chapter 3, which gives a theoretical and empirical analysis of the pattern of change of societal metabolism associated with the Jevons Paradox, to examine current literature and to provide an empirical macroeconomic analysis of various countries and regions to determine if the Jevons Paradox exists. Such an analysis is important, especially given the current debates on global warming and peak oil. People around the globe have to contend with problems related to pollution, mostly caused by energy consumption. While rare, energy blackouts are occurring more often as the demand for energy rises. Every day there are news reports on higher energy costs and a diminishing supply of natural resources to use for energy production. Yet we as world citizens are told not to worry by politicians, scientists, economists and other stakeholders, because technological advances will serve as a panacea to our problems. If the Jevons Paradox does exist, then technology as a liberator is a myth and appropriate sustainable development policies and behaviours need to be adapted before it is too late. This chapter seeks to contribute to the debate by illustrating whether or not the Jevons Paradox may exist for various countries and regions so that stakeholders can take action.

BACKGROUND

A detailed examination of the historical background of the Jevons Paradox was presented in Chapter 2, but this section will be used to refresh the reader's memory. In 1865 William Stanley Jevons eloquently detailed in his book *The Coal Question* that increased demand for a resource will occur as the result of improved efficiency in using that resource. Specifically, he explored the history of the steam engine to show how efficiency improvements lead to increases in the scale of production and, therefore, increase demand for coal. Jevons wrote:

> *Every such improvement of the engine, when affected, does but accelerate anew the consumption of coal. Every branch of manufacture*

receives a fresh impulse – hand labour is still further replaced by mechanical labour. (Jevons, 1865)

More than one hundred years later, Nicholas Georgescu-Roegen (1975) found that technological improvements tend to be energy-using and labour-saving, through the use of more powerful energy converters. In essence, the efficiency gains lead to an implicit decrease in prices, which causes greater demand because the same allocated budget can purchase a larger consumption bundle.

A considerable amount of research has been done on the Jevons Paradox, mostly under the title of the rebound effect. This research has examined a variety of uses and sectors. For example, Eiman Zein-Elabdin (1997) estimated charcoal supply and demand elasticities to calculate the rebound effects from more efficient stoves in the Sudan. The author found that the charcoal markets in the Sudan are characterized by low elasticities. He also calculated that 42 per cent of fuel savings are lost due to large price adjustments, since low elasticities put more of the burden of market adjustment on prices than on quantities. Lastly, Zein-Elabdin found that price-related effects may be small due to increases in charcoal prices.

Reinhard Haas and Lee Schipper (1998) extended the literature by investigating the function of efficiency for explaining and projecting energy demand in the residential sector. They found that technical efficiency is an important factor to consider for energy demand. They substantiated this claim by calculating near-zero price elasticities. Haas and Schipper also found that the absolute magnitude of both price elasticities and income elasticities over time may decrease, calling into question the assumption of constant elasticities. They used their findings to suggest that projecting energy demand using traditional constant and symmetric elasticity will lead to drastic overestimations.

Joyashree Roy (2000) built on these findings and explored the effect of efficiency gains on energy consumption in three sectors in India, namely an efficient lighting programme in rural households and efficiency effects in the industrial and transportation sectors. He found that, typical for developing countries, efficiency improvements lead to large rebound effects. In the rural residential sector, Roy calculated the short- and long-run elasticities for kerosene and liquid petroleum gas and found that demand will increase to a higher level than before. In the industrial sector, he calculated the energy price elasticity and found that with increased productivity and/or a price decrease, energy consumption similarly increases. Lastly, in the transportation sector he hypothesized that technological improvements will increase the demand for driving, thus producing a large rebound effect. These results are important, particularly for developing countries, because there tends to be a large amount of unmet demand in these nations.

Mark Jaccard and Chris Bataille (2000) complemented the literature by estimating elasticities of substitution for firms and households in Canada. They used simulation models to calculate the elasticity of substitution for capital and energy. Their results indicated that there is weak substitution between capital and energy, which suggested that the magnitude of the rebound effect is fairly low.

The literature was then extended by those interested in calculating how much of a rebound existed. For instance, Lee Schipper and Michael Grubb (2000) analysed data on energy use, energy intensity and prices in a variety of economic sectors for more than a dozen International Energy Agency countries to identify the rebound effects. They found a micro rebound elasticity of 5 per cent to 15 per cent and no substantial macro rebound in any sector or economy-wide. However, the authors caution that these results cannot be applied to developing countries, a very important exception. Additionally, the authors ignore the indirect rebounds which can be, and often are, very significant.

Reinhard Haas and Peter Biermayr (2000) investigated the effects of energy efficiency improvements on household energy consumption in Austria. They used time-series and cross-sectional analyses to estimate the size of the rebound effect. They estimated that the direct rebound effect from efficiency improvements leads to a 20 per cent to 30 per cent increase in energy consumption. Haas and Biermayr concluded that efficiency improvements will lead to a much smaller reduction in energy consumption and the related carbon dioxide emissions than policymakers believe.

Peter Berkhout, Jos Muskens and Jan Velthuijsen (2000) attempted to provide rigorous definitions of the rebound effect for both the single commodity case and the multiple commodity case. The authors found that the definition for a single commodity is different from that for multiple commodities. They developed a mathematical formulation of the rebound effect and use this to examine the case of The Netherlands. They found the rebound effect, under all definitions, to be between 0 per cent and 15 per cent. However, they did not provide the rigorous definitions they were seeking to supply in the paper.

Lorna Greening, David Greene and Carmen Difiglio (2000) conducted a survey of over 75 studies on the direct rebound effect. The authors restricted themselves to exploring the effects of fuel efficiency on specific energy services instead of on fuel consumption. They found that the size of the rebound was determined in large part by how the rebound effect was defined. They also found that the majority of studies focused on the residential sector. However, the authors found that these studies lack specification of residential fuel-consuming behaviour and recognition of capital attribute choice behaviour. Furthermore, the authors claimed that these studies suffer from omitted variable bias. The studies that Greening et al examined suggest a 0 per cent to 50 per cent rebound effect for a 100 per cent increase in energy efficiency in the residential sector. The industrial and commercial studies that the authors cited indicated that technical efficiency led to fuel savings which were only slightly reduced by increased demand. They concluded the paper stating that for the literature they reviewed for the US the rebound is not significant enough to mitigate the importance of energy efficiency as a tool to reduce carbon emissions, but that climate policies relying solely on energy-efficient technologies may need support from market and incentive mechanisms.

Grant Allan, Nick Hanley, Peter McGregor, Kim Swales and Karen Turner (2007) used an economy–energy–environment computable general equilibrium model for the UK to examine the impact of an economy-wide increase of 5 per cent in energy efficiency. They calculated rebound effects of 30 per cent to 50 per cent but no increase in energy use – in other words no 'backfire', where rebound is greater than 100 per cent. However, their assumptions regarding the structure of the labour market, production elasticities, time period and government consumption make their results very sensitive.

Runar Brannlund, Tarek Ghalwash and Jonas Nordstrom (2007) explored how the rebound effect affects the energy consumption of Swedish households and the related carbon dioxide, sulphur dioxide and nitrogen oxide emissions. They estimated the necessary change in taxation to keep carbon dioxide emissions at their initial levels and how this affected emissions of sulphur dioxide and nitrogen oxides. They found that an increase in energy efficiency of 20 per cent would increase carbon dioxide emissions by 5 per cent. Brannlund et al also calculated that to keep carbon dioxide emissions at their initial levels the carbon dioxide tax would have to be increased by 130 per cent. Furthermore, the tax would lead to a reduction in sulphur dioxide but an increase in nitrogen oxides. They concluded that if the marginal damages from sulphur dioxide and nitrogen oxides are non-constant then other policies must be adopted.

Harry D. Saunders (2000) explored the rebound effect from the macro side. He found that the evidence, or lack thereof, of a rebound in energy-intensity ratios can be obscured, in other words that energy intensity can increase with a rebound in existence. He also concluded that the degree of the rebound is dependent on the fuel price elasticity of supply and that the impact on gross domestic product will be small. He also found that fuel efficiency technologies that affect other non-fuel factors generate a large rebound effect. These results are important because they illustrate the significance of understanding the fuel price elasticity of substitution by policymakers developing environmental strategies.

Kenneth Small and Kurt Van Dender (2006) estimated the direct rebound effect for vehicles in the US from 1966 to 2001. Their model distinguished between autocorrelation and lagged effects, included a measure for fuel-economy standards, and permitted the rebound to vary according to income, urbanization and fuel costs. They calculated a short-run rebound of 4.5 per cent and a long-run rebound of 22.2 per cent. They also found that rising real income, assisted by falling fuel prices, caused the rebound to diminish substantially over the study period. Small and Van Dender also used a subset of their data, from 1997 to 2001, and calculated a short-run rebound of 2.2 per cent and a long-run rebound of 10.7 per cent. These results were considerably smaller than the values of 3.1 per cent and 15.3 per cent typically used for policy analysis.

Sverre Grepperud and Ingeborg Rasmussen (2004) applied a general equilibrium model to the Norwegian economy to examine whether energy efficiency improvements would lead to rebound effects, looking specifically at energy efficiency improvements in electricity and oil. Their results indicated that the rebound effect is very significant in the manufacturing sectors. However, in other sectors they found that the rebound effects are weak or nearly non-existent.

Peter de Haan, Michael Mueller and Anja Peters (2006) explored two types of rebound effects, an increase in car size and an increase in vehicle ownership, in relation to the introduction of hybrid cars into the Swiss automobile market. The authors conducted a survey of 367 buyers of the Toyota Prius 2 in Switzerland, with a response rate of 82.6 per cent. They collected data on the car that was replaced by the hybrid and all other cars owned by the household. De Haan et al also developed a model to describe the relationship between first-time car buyers, replacement vehicles and purchases of supplemental vehicles using the Swiss survey on mobility behaviour. Their results indicate that neither of the rebound effects could be confirmed. However, they did not examine the rebound effect for increased kilometres driven on an annual basis, which had been suggested to exist by previous research on the topic (Binswanger, 2001).

Others extended the literature by exploring the effects on the environment and what mechanisms exist to promote a reduction in energy consumption. For example, Fatih Birol and Jan Horst Keppler (2000) examined if and how much relative energy prices need to be changed in addition to energy efficiency improvements in order to meet Kyoto Protocol standards. They found that a change in price or efficiency will have a macro effect through the substitution of factors of production and goods, as well as an income effect. They also note that a change in relative prices and energy efficiency are complementary to each other.

John Laitner (2000) explored the rebound effect debate from a historical perspective as well as using a macroeconomic analysis of the US. Laitner was also interested in the effects of rebound on carbon emissions. He found that there is strong evidence that cost-effective energy efficiency improvements will reduce energy consumption. However, Laitner provided one caveat, namely that consumer preferences could diminish any energy savings. He concluded by stating that increased awareness about the connection between energy consumption, environmental degradation and more efficient products should minimize the changes in consumer preferences.

Richard Howarth, Brent Haddad and Bruce Paton (2000) reviewed two programmes sponsored by the US Environmental Protection Agency, Green Lights and Energy Star, which promote energy-efficient technology. Their article questioned the rebound effect and focused on the barriers to efficiency. They found that the Green Lights programme induced firms to invest in energy-reducing lighting systems. The authors also found that the Energy Star programme led suppliers of computer and electronic equipment to significantly improve the energy efficiency of their products, which provide cost savings for equipment users. They argued that the achievements of these programmes are based on their success in reducing market failures stemming from imperfect information and from bounded rationality constraints that diminish the effectiveness of intra-firm organization and the coordination between equipment suppliers and their customers. They concluded by illustrating that the programmes will have little effect on the demand for energy, so energy efficiency improvements should lead to one-to-one reductions in energy consumption.

Mathias Binswanger (2001) examined the traditional neoclassical framework of a partial equilibrium analysis on the demand of one particular energy service. Binswanger went beyond the neoclassical model to explore a variety of possible causes of rebound to illustrate its relevance for ecological economics. He found that the rebound effect with respect to households is often neglected in the sustainable development debate. Furthermore, Binswanger found that some sectors may exhibit energy savings but the household sector will demand more of the products from these sectors, which will result in an overall increase in energy consumption. Binswanger concluded by illustrating how the traditional neoclassical single-service model does not capture all the effects that result in a rebound effect and is therefore not effective in developing sustainable development policies.

Vincent Otto, Andreas Loschel and John Reilly (2006) developed a forward-looking computable general equilibrium model to capture the empirical connections between carbon dioxide emissions associated with energy use, efficiency improvements and the economy. They found that cost-effective climate policy should include a combination of research and development subsidies aimed at reducing carbon dioxide emissions and carbon dioxide constraints, even though research and development subsidies increase the shadow price of the carbon dioxide price due to a large rebound from stimulating innovation. Their results indicated that carbon dioxide constraints aimed at carbon-dioxide-intensive sectors are more cost-effective than a uniform carbon dioxide policy. They reached this conclusion because they found that differentiated carbon dioxide prices encourage growth in non-carbon-dioxide-intensive sectors, making it cost-effective for carbon-dioxide-intensive sectors to bear more of the abatement burden.

This brief examination of some of the current literature illustrates the wide disparity of the estimates of the rebound effect and just how difficult a task it is to determine if the Jevons Paradox exists only for a specific sector or if the phenomenon is economy-wide. The following sections of this chapter extend these studies by exploring which factors lead to the Jevons Paradox on an economy-wide scale.

A JEVONS PARADOX MODELLING APPROACH

The Jevons Paradox has important implications for environmental conditions; therefore the analysis presented in this chapter takes a different approach from the research presented in the previous section because the environment (where energy supplies come from) is a complex system. A reduction in resource consumption through increased energy efficiency would, in theory, be good for complex systems because less pollution is released. However, efficiency gains would only result in less pollution if the complex adaptive system were not able to adapt. Unfortunately, complex systems tend to adapt quickly, and once technological improvements are introduced one or both of the following can occur: (1) an expansion of current levels of activity within the original setting and (2) an increase in the option space with additional activities (Giampietro and Mayumi, 2005).

Increases in energy efficiency lower the consumption of inputs, which in turn lowers the price of production. As the price of production declines, demand and consumption increase, resulting in the Jevons Paradox. Moreover, improvements in efficiency are transformed into new and more complex behaviour outside the system (Giampietro and Mayumi, 2005). As a resource becomes more efficient to use, and more affordable, current technology will be used more or new technology will be introduced that contains more options and features.

As shown previously, examples of the Jevons Paradox are numerous. However, the increase in demand for a resource is not strictly confined to products that use that resource more efficiently; it can also involve other end-uses because they compete for the same overall budget (Khazzoom, 1980). Therefore, not only does a direct micro rebound effect exist, but there is also an indirect macro rebound. In the case of a macro rebound, there is an income effect causing an increase in real income which permits the consumer to purchase an upgrade in quality, as well as an increase in demand (Wirl, 1997, pp20, 26–27, 31, 41 and 197; Saunders, 2000). Thus the rebound effects are economy-wide, not specific to just one sector, product or end-use, whereas a micro-level effect is specific to one product, sector or end-use. This chapter builds upon these findings by describing data and models that are then used to empirically examine if there is any suggestion that the Jevons Paradox exists in a variety of case studies.

The approaches taken in the research described previously tend to focus on specific energy uses and/or calculating elasticities of demand and substitution to estimate *direct* rebound effects. These previous studies did not attempt to determine which factor or factors caused the rebound. Rather, they explored only a small percentage of the change in total energy consumption caused by very specific technological efficiency improvements. However, energy efficiency improvements tend to be economy-wide or to affect the macro economy, making it vital to explore the total rebound for a country and which parameters have the greatest impact on energy demand. Therefore, a macro-level analysis is provided

Table 4.1   *Mapping of variables to the I = PAT equation*

Environmental impact (I)	Population (P)	Affluence (A)	Technology (T)
Total primary energy consumption	Population density	Gross domestic product (GDP)	Energy intensity
	Population	Exports	
	Urban population	Imports	
	Rural population	Household consumption	
		Government consumption	

in this chapter with the specific aim of obtaining a complete understanding of the relationship between energy consumption and energy efficiency so that policymakers can have an economy-wide understanding of the Jevons Paradox when developing energy policy.

To perform this macro-level analysis, the I = PAT model developed by Paul Ehrlich and John Holdren (1971) will be used to determine the key variables in the models presented in this chapter. The I = PAT equation illustrates the impact of population (P), affluence (A) and technology (T) on the environment. Therefore, for our purposes here the left hand side of the equation, environmental impact (I), will correspond to total primary energy consumption. The right hand side of the equation requires examination of the three main macro factors that influence energy consumption:

1   total population, population density, urban population and rural population (P);
2   gross domestic product and its individual components (except for investment) (A); and
3   energy intensity as BTUs (British thermal units) per GDP, a proxy for technological improvements (T).

Mapping the variables in this manner allows one to see which variables have the greatest impact on energy consumption (environmental impact). Table 4.1 presents the mapping of the variables used for analysis to the I = PAT equation.

DESCRIPTION OF VARIABLES AND MODELS

Total primary energy consumption is defined by the International Energy Agency as all energy that is consumed by end-users plus losses that occur in the generation, transmission and distribution. This type of energy consumption is employed as the measure of energy usage. Many stakeholders believe that population is the main reason for increased energy consumption. Therefore, total population is used as a covariate to test this hypothesis. Population density is also used. This variable serves as a proxy for urbanization and access to energy, particularly in developing countries, which contributes to increased energy consumption. Access to energy increases with the size of the urban area, and this access is a major catalyst for people migrating to cities. In some of the case studies presented below, urban and rural population is used. By dividing the population into urban and rural, one can see if there are differences in energy consumption between the two regions, and if there is a pattern of migration. GDP, measured in constant 2000 US$, is used as a measure of economic activity. GDP in constant 2000 international $ (purchasing power parity) is also used to correct for any differences in exchange or inflation rates between countries. The individual components of GDP – household consumption, government consumption, imports and exports, all measured in constant 2000 US$ – are also used as a measure of economic activity and consumption of goods and services on a more detailed level. Household consumption is used to determine if the products that households are purchasing consume a lot of energy. Government consumption is explored to determine if government purchases are contributing to energy consumption or not. Exports and imports are used because international trade is an important component of most economies. In fact, many large economies, particularly in developing countries, import so they can export. Energy intensity, measured as BTU per constant 2000 US$ and in constant 2000 international $, is used as a proxy for technological improvements that are energy efficiency improvements. Each of these variables is used in different configurations of models in the case studies.

Two different modelling approaches are used in the case studies in the sections below. The first method, ordinary least squares (OLS) models, are used to analyse the data for the US and Brazil case studies. In all of the models for both of these cases the Durbin-Watson statistics were less than the lower critical values at the 5 per cent significance level, indicating that first-order correlation was present among the disturbances. Therefore, a GARCH(1,1) model was used to analyse the time-series data. Time-series data allows for a sequence of observations to be examined to predict the future values of the variables. However, time-series data is likely to have heteroscedasticity and autocorrelation. The GARCH(1,1) model, an alternative to standard time-series processes, was used to correct for these problems, imposing a special structure on heteroscedastic disturbances to obtain OLS best linear unbiased estimators (Murray, 2006). Maximum likelihood estimation of the GARCH(1,1) model was used to determine if autocorrelation was present and to obtain estimators that are unbiased and error terms that are randomly distributed.

In the second approach, time-series cross-sectional (TSCS) regression models are used for analysis for the European and Asian regional case studies. TSCS data creates variability, thus eliminating heterogeneity, and provides more informative results by eliminating the need for lengthy time-series by utilizing the information available on the dynamic reactions of each subject (Kennedy, 2003). Furthermore, TSCS data permits both spatial and temporal effects to be examined, allowing a subject, in our case energy consumption, to be studied over multiple sites and observed over a defined timeframe. Using time-series with cross-sections can only enhance the quality and quantity of data, which would be impossible using only one of these two dimensions (Gujarati, 2003).

Therefore, TSCS is an invaluable tool. However, the regression estimates are still likely to be biased and inefficient. Several problems are frequent in TSCS analyses. First, errors tend to be serially correlated because observations and traits that characterize the error term tend to be interdependent across time. Second, the error terms tend to be correlated across countries. Third, heteroscedasticity is likely in TSCS data sets because the error variances tend not to be constant across countries. Fourth, the error terms may contain both spatial and temporal effects that produce a regression model with heteroscedastic and autocorrelated errors. The fifth and final problem that arises with TSCS analyses is that errors tend to reflect partial causal heterogeneity across time, space or both (Hicks, 1994).

In addition to the problems listed above, correlation in the data set is expected. To correct for these problems, maximum likelihood estimators are calculated by iterating the generalized least squares method to correct for group-wise heteroscedasticity and correlation across groups, as well as group-specific autocorrelation. Furthermore, if, as expected, correlation is present in the variables chosen for this study, this technique provides unbiased estimators (Greene, 2000). It is important to note that this technique does not produce a goodness-of-fit measure.

Variations of two generalized models, Equations 1 and 2, are used for analysis in both the GARCH(1,1) and TSCS approaches. These models were chosen to capture as many macro-level effects that may be present as possible, enabling some insight to be obtained on the energy policies of the various countries and regions in the case studies and what may be the best way to proceed on any future energy policies. Unfortunately, there is little, if any, work like this at national or regional levels for the Jevons Paradox (Polimeni and Polimeni, 2006, 2007a and 2007b; Polimeni 2007). Thus the information obtained from these models will provide an important analysis of the existing energy policies in the countries and regions examined and how well they are performing.

The first generalized model estimates energy consumption in relation to energy intensity, gross domestic product, and either population or population density. This model is estimated with GDP and energy intensity measured in both constant 2000 US\$ and in purchasing power parity terms.

$$EC = \beta_1 + \beta_2 EI + \beta_3 GDP + \beta_4 P \qquad (1)$$

where:

EC = total primary energy consumption;
EI = energy intensity;
GDP = gross domestic product; and
P = total population or population density.

This model will provide supporting evidence for the macro level manifestation of the Jevons Paradox if the coefficients for energy consumption and energy intensity are positive, indicating that energy consumption should decrease as energy intensity decreases. The coefficients for both total primary energy consumption and energy intensity are expected to be positive. Additionally, the coefficients for GDP and population or population density are expected also to be positive – as GDP increases so will energy consumption, and as population or population density increases so will energy consumption because there are more people. However, if the marginal rate of energy intensity has the greatest magnitude of the covariates on energy consumption, then there is strong evidence that the Jevons Paradox may exist. If energy intensity does not have the marginal rate with the greatest impact on energy consumption then the Jevons Paradox may be obscured or non-existent as other variables, as many stakeholders believe, may be responsible for the increased use of energy.

The second generalized model estimates energy consumption as a function of household consumption, government consumption, imports, exports, energy intensity, and either population or population density. All of these variables are measured in constant 2000 US\$. In this model, GDP is disaggregated into its individual parts (household consumption, government consumption, imports and exports) to determine which of the individual components impacts energy consumption the most. Furthermore, imports and exports are included to determine if international trade has an impact on energy consumption.

$$EC = \beta_1 + \beta_2 EI + \beta_3 HH + \beta_4 G + \beta_5 X + \beta_6 M + \beta_7 P \qquad (2)$$

where:

EC = total primary energy consumption;
EI = energy intensity;
HH = household consumption;
G = government consumption;
X = exports;
M = imports; and
P = total population or population density.

The signs of the coefficients for each of the variables in the second general model are expected to be positive, with the exception of imports. The coefficient for imports is expected to be either positive or negative.

Both generalized models are intended to specifically determine which variables have an impact on energy consumption. Furthermore, each of the models is used to provide some statistical evidence that the Jevons Paradox may exist at the macro level and to eliminate covariates that have little, if any, impact. Therefore, the models are used to determine which factors, if there is some empirical evidence that the Jevons Paradox may exist at the macro level, drive the increased energy consumption. The results and findings of these models are discussed below.

The US

The first study that will be examined is the case of the US. As one of the leading carbon dioxide, sulphur dioxide and nitrous oxide polluters in the world, the global environmental implications of the possible existence of the Jevons Paradox in the US are numerous. As illustrated in Figure 4.1 energy consumption in the US increased by nearly 100 per cent from 1960 to 2004. This increase corresponds to nearly 31 quadrillion BTU.

At the same time as this substantial increase in energy consumption, the US experienced a technological boom, with energy intensity decreasing by 113 per cent during the 45-year period (Figure 4.2). Most of this decrease in energy intensity occurred after 1970, as energy intensity remained nearly constant in the previous decade.

A quick examination of the data would lead one to find that the Jevons Paradox does exist for the US during the 1960 to 2004 time period. However, deeper statistical analysis is needed to get a better understanding of which variables may affect total primary energy consumption the most. To have a complete and consistent data set, data for the 1975 to 2004 time period is used for analysis. This

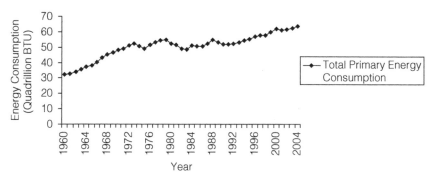

Figure 4.1   *Energy consumption in the US, 1960–2004*

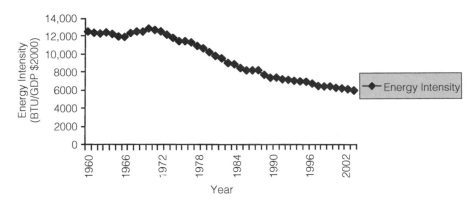

Figure 4.2   *Energy intensity in the US, 1960–2004*

data was obtained from the Energy Information Administration's *International Energy Annual* and the World Bank.

As explained in the previous section, a GARCH(1,1) modelling approach was used to analyse the data. Three models developed from the two generalized models described above are used in the GARCH(1,1) approach; the results of which are shown in Table 4.2. In all the models presented, the first-order autoregressive parameter ρ (rho) for the regression was significant at the 95 per cent confidence level, indicating that the GARCH(1,1) model fits the data significantly better than the corresponding OLS model because of autocorrelation. Therefore, only the GARCH(1,1) results are presented.

The first model examined the impacts of GDP, population and energy intensity on total primary energy consumption for the 1975 to 2004 time period. Of particular interest, population and population density are not significant in this model even though both population and population density increased by approximately 36 per cent over the study period. However, both GDP, measured in constant 2000 US$, and energy intensity were significant. During the study period GDP increased nearly 152 per cent and energy intensity decreased almost 50 per cent. The coefficient for GDP is both significant and positive, indicating that as GDP increases so will total primary energy consumption. This result, as outlined previously, is expected. The coefficient for energy intensity is also significant and positive. To make this result more intuitive, as energy intensity decreases, which occurs in most economies, energy consumption should decrease. However, as depicted in Figures 4.1 and 4.2, this is not the case for the US. Also of note is the size of the magnitudes of the two variables. The results indicate that energy intensity, on a percentage basis, has a greater impact on energy consumption than GDP. However, in absolute terms, GDP has the greater impact because GDP is measured in trillions of dollars. The second model is identical to the first except

Table 4.2   *Regression results for the US*

Variable	Model 1	Model 2	Model 3
Constant	−32.9998 (8.3193) [0.0001]	38.747 (3.6577) [0.0000]	−21.8296 (11.0992) [0.0492]
GDP (constant 2000 $ US)	0.000000000006 (0.0000) [0.0000]		
GDP (constant 2000 $ international)		0.000000000002 (0.0000) [0.0000]	
Household consumption (constant 2000 $ US)			0.000000000008 (0.0000) [0.0000]
Energy intensity (BTU/ constant 2000 $ US)	0.0049 (0.0006) [0.0000]		0.0044 (0.0008) [0.0000]
Energy intensity (BTU/ constant 2000 $ international)		0.000321 (0.0001) [0.0165]	
Rho	0.96619 (0.0479) [0.0000]	0.65217 (0.1408) [0.0000]	0.9403 (0.0632) [0.0000]

(Standard errors reported in parentheses)
[p-values presented in square brackets]

that GDP and energy intensity are measured in purchasing power parity terms. As in the first model, population and population density are not significant. However, both GDP and energy intensity are significant, both with positive coefficients. The results show that there is a difference from the first model. In the second model, GDP decreased in magnitude by a third, while the magnitude of energy intensity decreased by a magnitude of fifteen. Like the first model, the magnitude of energy intensity is much greater in marginal terms. The results of the first two models suggest that the Jevons Paradox may exist for the US.

The third model examines the main components of GDP to determine which sector or sectors are driving the increase in energy consumption. As with the first two models, population and population density were included in the models; however, neither was significant. Additionally, government consumption, imports and exports were also found not to be significant. Therefore, the variables that were significant are household consumption and energy intensity. Both variables have positive coefficients, as expected, indicating that they cause increased energy consumption. The magnitudes of the coefficients of the variables are nearly identical to those in the first model. These findings indicate that household consumption is the main component of GDP that drives increasing energy consumption. Furthermore, the results also indicate that, from a marginal perspective, energy intensity is the factor that increases energy consumption the most.

Europe

The next study examines the case of Europe. Sixteen European countries (Austria, Belgium, Denmark, Finland, France, Germany, Greece, Ireland, Italy, The Netherlands, Norway, Portugal, Spain, Sweden, Switzerland and the UK) are included in this case study to explore if the Jevons Paradox may exist there. Other European countries were not included, such as those in Eastern Europe, either because a complete data set was not available or because their economies were very small in relation to the others in the region (in the case of Luxembourg, for example).

Internationally recognized as a leader in environmental matters, if the empirical evidence shows that the Jevons Paradox may be in existence for Europe then decision makers will have to readjust their environmental policies. Furthermore, if the leader in environmental issues exhibits the Jevons Paradox, then how much worse is the situation in other regions? Therefore, the results of this case study are particularly important.

In this case study, time-series cross-sectional (TSCS) regression models are used for analysis. The two generalized models described in Equations 1 and 2 are used in the TSCS analysis. The same variables used in the first case study are also used here. The only difference is that rural population and urban population are included in one variation of Equation 1. The variables are analysed for the 1980–2004 time period, with data obtained from the Energy Information Administration's *International Energy Annual* and the World Bank.

Figure 4.3 illustrates energy consumption for the European countries in the case study. Every country experienced increased energy consumption during this time period except for Germany, which had a very minor decrease. Therefore, Europe, as defined by the selected sixteen countries, saw an overall increase in energy consumption.

At the same time that the region saw an increase in energy consumption, energy intensity during the 25-year time period decreased. The only countries that did not experience a decrease in energy intensity were Greece, Portugal and Spain. These results are shown in Figure 4.4.

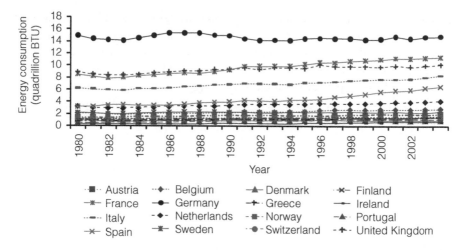

Figure 4.3   *Energy consumption in 16 selected European countries,*
1980–2004

Therefore, like the previous case study, a cursory examination of the data would lead one to believe that the Jevons Paradox exists in Europe. However, deeper statistical analysis is needed to confirm this. The results of seven TSCS models will be presented to obtain a better understanding of which variables are influencing energy consumption in these 16 European countries the most.

The results of the first five models are presented in Table 4.3. The first model tests the relationship between total primary energy consumption and population, GDP in constant 2000 US$ and energy intensity in BTU per constant 2000 US$. Each of the variables in the model is significant and positive. The results indicate that population in the study countries is the primary factor for the

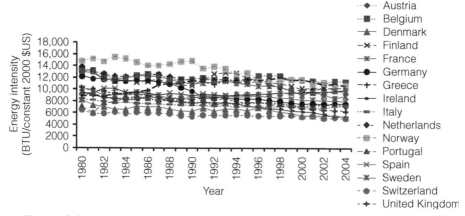

Figure 4.4   *Energy intensity in 16 selected European countries, 1980–2004*

Table 4.3   *Regression results for 16 selected European countries: Models 1–5*

Variable	Model 1	Model 2	Model 3	Model 4	Model 5
Constant	-0.8791 (0.1022) [0.0000]	-0.6836 (0.0155) [0.0000]	-1.0055 (0.0125) [0.0000]	-0.8977 (0.0165) [0.0000]	-0.8216 (0.0097) [0.0000]
Population	0.02893 (0.0011) [0.0000]		0.0165 (0.0007) [0.0000]		0.015545 (0.0012) [0.0000]
Population density (people per square kilometre)		-0.0012 (0.0001) [0.0000]		-0.0002 (0.0001) [0.0000]	
GDP (constant 2000 $ US)	0.000000000007 (0.0000) [0.0000]	0.0000000000086 (0.0000) [0.0000]			
GDP (constant 2000 $ international)			0.000000000006 (0.0000) [0.0000]	0.000000000006 (0.0000) [0.0000]	
Exports (constant 2000 $ US)					0.000000000004 (0.0000) [0.0000]
Imports (constant 2000 $ US)					-0.0000000000034 (0.0000) [0.0000]
Household consumption (constant 2000 $ US)					0.00000000001 (0.0000) [0.0000]
Government consumption (constant 2000 $ US)					0.000000000007 (0.0000) [0.0000]
Energy intensity (BTU/constant 2000 $ US)	0.0001 (0.0000) [0.0000]	0.0001 (0.0000) [0.0000]			0.0001 (0.0000) [0.0000]
Energy intensity (BTU/constant 2000 $ international)			0.0001 (0.0000) [0.0000]	0.0001 (0.0000) [0.0000]	

(Standard errors reported in parentheses)
[p-values presented in square brackets]

increased energy consumption in the region. This result is somewhat surprising given the low birth-rates the study countries have had in the past 25 years. Examining the data leads one to conclude that much of this result is due to two of the major economies in the region, the UK and France. While the magnitude of the coefficient for energy intensity is considerably less than that for population, the outcome does suggest that the Jevons Paradox may be in existence. The second model examines the same relationship as the first with the only difference being that purchasing power parity terms are used. The results are very similar to those in the first model and, as such, will not be commented on further here.

The third model explores the relationship between energy consumption and population density, GDP and energy intensity. The coefficient for population density is significant and negative. This result suggests that as the populace of these 16 countries migrates to urban areas, energy consumption is reduced. This result could possibly be due to decreased energy consumption in the transportation sector. Moreover, the finding also suggests that agriculture in the study region uses a lot of energy. Both GDP and energy intensity are significant and with positive coefficients. However, the marginal rate for energy intensity is much greater than that for GDP, further suggesting that the Jevons Paradox may exist for the region. Model 4 examined the same relationship as Model 3, but measures GDP and energy intensity in purchasing power parity terms. As in the first model, the results from this model using purchasing power parity data are nearly identical to the results in the third model. Therefore, no additional analysis will be provided. The fifth model extends the analysis of Models 3 and 4 by examining the relationship between energy consumption and urban population, rural population, GDP in constant 2000 US$ and energy intensity in BTU per constant 2000 US$. Recall that population density had a negative coefficient. Model 5 enables us to test this relationship further by exploring which population, urban or rural, has a larger impact on energy consumption. Both variables are significant and have positive coefficients. Interestingly enough, the coefficient for rural population is four times greater than the coefficient for urban population, suggesting that agricultural production in the region consumes a lot of energy. The other interesting result is that energy intensity is positive and has the coefficient with the greatest magnitude. This result means that on a marginal basis energy intensity has the greatest impact on energy consumption in this model, again indicating that the Jevons Paradox may exist.

The sixth and seventh models, presented in Table 4.4, extend the analysis further, dissecting GDP into its main components. Model 6 explores the relationship between total primary energy consumption and population, exports, imports, household consumption, government consumption and energy intensity; Model 7 examines the same relationship except that population density is substituted for population. The results of both models confirm the findings from the previous models. Once again, population has the greatest marginal impact on energy consumption while population density reduces energy consumption. These results follow the analysis that was provided for the first five models. Furthermore, the coefficient for energy intensity is positive and significant in each of the models, suggesting that the Jevons Paradox is, at worst, partially occurring. However, the main benefit of these two models is to obtain information on how the components of GDP affect energy consumption within the 16 study countries. Of particular interest are the results for the trade variables. The exports variable is positive, indicating that the 16 European countries are exporting products which require a large amount of energy consumption during the production process. On the other hand, the imports variable has a negative coefficient, implying that these countries are importing products that consume less energy than the products the populace are currently using, such as clothing, oil and food. This result suggests that the 16 countries

Table 4.4   *Regression results for 16 selected European countries: Models 6–7*

Variable	Model 6	Model 7
Constant	−0.3281 (0.0130) [0.0000]	−0.9085 (0.0095) [0.0000]
Population density (people per square kilometre)	−0.00389 (0.0001) [0.0000]	
Urban population		0.00000001 (0.0000) [0.0000]
Rural population		0.00000004 (0.0000) [0.0000]
GDP (constant 2000 $ US)		0.000000000008 (0.0000) [0.0000]
Exports (constant 2000 $ US)	0.0000000000048 (0.0000) [0.0000]	
Imports (constant 2000 $ US)	−0.0000000000046 (0.0000) [0.0000]	
Household consumption (constant 2000 $ US)	0.000000000012 (0.0000) [0.0000]	
Government consumption (constant 2000 $ US)	0.000000000005 (0.0000) [0.0000]	
Energy intensity (BTU/ constant 2000 $ US)	0.0001 (0.0000) [0.0000]	0.0001 (0.0000) [0.0000]

(Standard errors reported in parentheses)
[p-values presented in square brackets]

as a whole may be exporting their pollution (resulting from energy intensive production) to less developed countries. If true, the impact on the environment in developing countries in particular will be negative, further creating disparity between developed and developing countries. The other components of GDP, household and government consumption are both significant, with positive coefficients. As anticipated, as these variables increase energy consumption does as well.

Asia

The third case study looks at the Asian region. In this case study 12 of the largest economies in Asia are investigated. These countries are Australia, Bangladesh, China, Hong Kong, India, Indonesia, Japan, South Korea, Malaysia, New Zealand, the Philippines and Thailand. As in the European case study, TSCS analysis will be used to study the Asian region. Other countries, such as Vietnam and Singapore, would have been included, but complete data sets for the other countries in the region were unavailable. The countries selected in this region vary considerably, ranging from developing nations to those on the verge of developed nation status to fully developed status, providing a nice comparison to the other case studies analysed in this chapter.

As illustrated in Figure 4.5, each of the countries included in the case study experienced an increase in energy consumption from 1980 to 2004. Therefore, we can conclude that the Asian region, as defined in this case study, exhibited an increase in energy consumption.

At the same time that the region was consuming more energy, the largest economies were going through a period of improved energy efficiency. However, seven of the countries did experience a slight increase in energy intensity, as

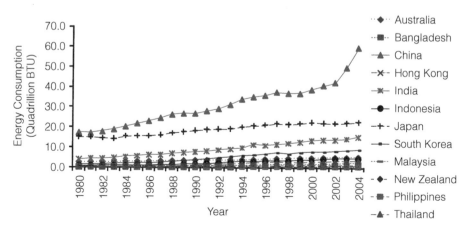

Figure 4.5   *Energy consumption in 12 selected Asian countries, 1980–2004*

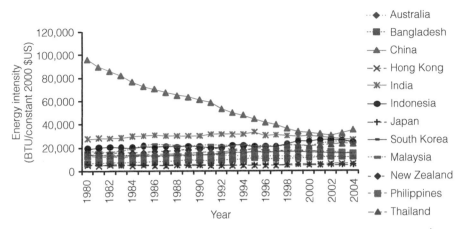

Figure 4.6 *Energy intensity in 12 selected Asian countries, 1980–2004*

shown in Figure 4.6: Bangladesh, Hong Kong, Indonesia, South Korea, Malaysia, the Philippines and Thailand.

The regression results for this case study are presented in Table 4.5. Equations 1 and 2 are used to develop six TSCS models for the region, similar to those used in the analysis of the European region. The first model investigates the relationship between energy consumption and population, GDP (measured in constant 2000 US$) and energy intensity (measured in BTU/constant 2000 US$). Each of the variables in the model is significant with positive coefficients. This finding indicates that total primary energy consumption will increase as each of the variables in the model increase. Of particular interest for our purposes here, the marginal rate of increase and the absolute magnitude of the coefficient for energy intensity is the largest of any of the variables, suggesting that the Jevons Paradox may be in existence in the Asian region as well. The second model examined the same relationship, only using GDP and energy intensity measured in purchasing power parity terms. The results were much the same, with only the one slight difference that the coefficient for energy intensity became greater in magnitude. This result indicates, at least in this particular model configuration, that the Jevons Paradox may be in existence for the study region.

The third and fourth models used the same configuration as in Models 1 and 2, but used population density instead of population. Model 3 used GDP and energy intensity measured in constant 2000 US$. The results indicate that population density has the greatest marginal impact on energy consumption. However, when GDP and energy intensity are measured in purchasing power parity terms, the coefficient for population density becomes negative. This finding indicates that the currency in these 12 countries is weak and that rural areas consume a lot of energy in comparison to their urban counterparts.

Table 4.5 *Regression results for the Asian region case study*

Variable	Model 1	Model 2	Model 3	Model 4	Model 5	Model 6
Constant	-0.53626 (0.0069) [0.0000]	-0.4169 (0.0194) [0.0000]	-1.014166 (0.0346) [0.0000]	-0.3117 (0.0203) [0.0000]	-0.3376 (0.0238) [0.0000]	-0.1266 (0.0228) [0.0000]
Population	0.00000000001 (0.0000) [0.0000]		0.0000000026 (0.0000) [0.0000]		0.0000000023 (0.0000) [0.0000]	
Population density (people per square kilometre)		0.0001 (0.0000) [0.0000]		-0.0001 (0.0000) [0.0000]		-0.000054 (0.0000) [0.0000]
GDP (constant 2000 $ US)	0.0000000000001 (0.0000) [0.0000]	0.0000000000001 (0.0000) [0.0000]				
GDP (constant 2000 $ international)			0.0000000000047 (0.0000) [0.0000]	0.0000000000043 (0.0000) [0.0000]		
Exports (constant 2000 $ US)					0.000000000000042 (0.0000) [0.0000]	0.000000000000012 (0.0000) [0.0292]
Imports (constant 2000 $ US)					-0.000000000000026 (0.0000) [0.0000]	0.000000000000012 (0.0000) [0.0289]

Household consumption (constant 2000 $ US)			0.00000000002 (0.0000) [0.0007]	0.0000000000055 (0.0000) [0.0000]
Government Consumption (constant 2000 $ US)			0.00000000000028 (0.0000) [0.0000]	0.00000000000013 (0.0000) [0.0000]
Energy intensity (BTU/ constant 2000 $ US)	0.000045 (0.0000) [0.0000]	0.00004 (0.0000) [0.0000]	0.000042 (0.0000) [0.0000]	0.000029 (0.0000) [0.0000]
Energy intensity (BTU/ constant 2000 $ international)	0.0001 (0.0000) [0.0000]	0.0001 (0.0000) [0.0000]		

(Standard errors reported in parentheses)
[p-values presented in square brackets]

Moreover, the result indicates that rural energy consumption is largely due to agricultural production methods.

The fifth and sixth models build on the results of the first four models to find what components of GDP influence energy consumption the most. The fifth model explores the relationship between energy consumption and population, exports, imports, household consumption, government consumption and energy intensity. All the variables in the model are significant and all the variables have positive coefficients except for imports. The sixth model is the same as the fifth model, with the only difference that population density is used instead of population. Once again, all the variables are significant. However, in this model all the coefficients of the variables are positive except for population density. The results of these two models are interesting because the sign of the coefficient for imports changed from negative in Model 5 to positive in Model 6. Furthermore, the results build upon those in the previous four models. The change of sign of the coefficient for imports, along with the negative coefficient for population density, suggests that a large percentage of the population in the study region is located in rural areas. The coefficient for imports in Model 5, which uses total population, indicates that the agricultural sector consumes a large amount of energy and that imports to the region use less energy than those that are produced domestically. However, the positive coefficient for imports in Model 6 suggests that urban areas are importing energy-consuming goods. This result also implies that urban areas in Asia contain heavy industry, producing goods such as textiles and consumer electronics. A final explanation for the results in Models 5 and 6 is that energy consumption from imports in the fifth model is spread out over the entire population, whereas in the sixth model energy population density serves as a proxy for urbanization.

Brazil

The last case study presented in this chapter will be Brazil. Brazil is an interesting case study because the country, as shown in Figures 4.7 and 4.8, has experienced an increase in both energy consumption and energy intensity from 1980 to 2004. This experience is different from the other case studies in this chapter, but provides meaningful insight nonetheless. Five models of various configurations are used to obtain an understanding of which factors are causing the increase in energy consumption.

The regression results of the models for Brazil are shown in Table 4.6. In all the models presented, the first-order autoregressive parameter ρ (rho) for the regression was significant at the 95 per cent confidence level, indicating that the GARCH(1,1) model fits the data significantly better than the corresponding OLS model because of autocorrelation. Therefore, like in the US case study only the GARCH(1,1) results are provided.

The first model explores the relationship between energy consumption and GDP, energy intensity and population. Population, however, is not significant. Therefore, the model consists of GDP and energy intensity as the covariates. Both

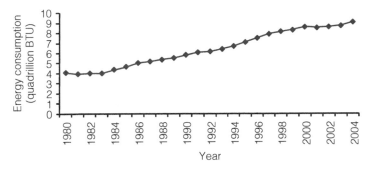

Figure 4.7   *Energy consumption in Brazil, 1980–2004*

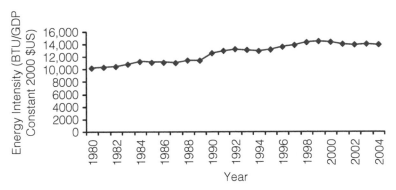

Figure 4.8   *Energy intensity in Brazil, 1980–2004*

variables are significant and have positive coefficients. However, the coefficient for energy intensity is much greater in magnitude than the coefficient for GDP, indicating that on a marginal level energy-efficiency technological improvements are more responsible for energy consumption. The second model examines the same relationship, with the only difference that the variables are measured in purchasing power parity terms. As with the first model, neither population nor population density is significant. The results in Model 2 are similar to those in Model 1, therefore no additional analysis will be provided for this model.

The third model inspects how total primary energy consumption is affected by exports, imports, household consumption, government consumption and energy intensity. Population and population density were not significant in this model either. The results of the model show that each of the variables is significant and has positive coefficients, implying that as these variables increase they cause energy consumption to increase. The results indicate that energy intensity has the greatest impact on energy consumption on a marginal basis. The results for the other variables suggest that the components of GDP are nearly equal in causing an increase in energy consumption.

The fourth and fifth models explore whether splitting population into rural and urban populations will shed some light on the effect of population on energy

Table 4.6 *Regression results for the Brazilian case study*

Variable	Model 1	Model 2	Model 3	Model 4	Model 5
Constant	-5.9164 (0.2677) [0.0000]	-5.9164 (0.2677) [0.0000]	-3.0335 (0.4395) [0.0000]	-0.3996 (0.9090) [0.6603]	6.5256 (1.0252) [0.0000]
Urban population				-0.000000012 (0.0000) [0.0002]	
Rural population				-0.0000001 (0.0000) [0.0000]	-0.0000002 (0.0000) [0.0000]
GDP (constant 2000 $ US)	0.00000000001 (0.0000) [0.0000]				
GDP (constant 2000 $ international)		0.0000000000059 (0.0000) [0.0000]		0.00000000012 (0.0000) [0.0000]	
Exports (constant 2000 $ US)			0.00000000019 (0.0000) [0.0000]		
Imports (constant 2000 $ US)			0.0000000001 (0.0000) [0.0101]		0.000000000068 (0.0000) [0.0217]

Household consumption (constant 2000 $ US)			0.000000000000082 (0.0000) [0.0000]		0.0000000000000079 (0.0000) [0.0000]
Government consumption (constant 2000 $ US)			0.000000000000014 (0.0000) [0.0000]		0.000000000000014 (0.0000) [0.0000]
Energy intensity (BTU/constant 2000 $ US)	0.00049 (0.0000) [0.0000]		0.0003 (0.0000) [0.0000]	0.0005 (0.0000) [0.0000]	0.0003 (0.0000) [0.0000]
Energy intensity (BTU/constant 2000 $ international)		0.0010 (0.0000) [0.0000]			
Rho	0.9496 (0.0640) [0.0000]	0.9496 (0.0640) [0.0000]	0.5542 (0.1699) [0.0011]	0.7798 (0.1278) [0.0000]	0.3211 (0.1933) [0.0967]

(Standard errors reported in parentheses)
[p-values presented in square brackets]

consumption. All the variables – rural population, urban population, GDP and energy intensity – are significant. However, the results are somewhat surprising. The coefficients for both urban and rural population are negative, implying that as these variables increase energy consumption will decrease. A deeper examination finds that the rural population in Brazil has decreased in the 25-year study period, thus explaining the negative coefficient. However, urban population during this time period increased, indicating that Brazilians are migrating to cities. However, this migration does not cause an increase in energy consumption. On the other hand, GDP and energy intensity both had positive coefficients, suggesting that as these variables increase, so will energy consumption. Most important, the coefficient for energy intensity was largest in magnitude, signifying that on a marginal basis energy efficiency actually causes energy consumption.

Model 5 examines the relationship between total primary energy consumption and urban population, rural population, exports, imports, household consumption, government consumption and energy intensity. All the variables in the model, except for exports and urban population, are significant. Furthermore, the significant variables all have positive coefficients except for rural population. The cause for the negative coefficient for rural population was explained in the analysis of Model 4 and will not be readdressed here. As in Model 3, the coefficients for the individual components of GDP are nearly identical, suggesting that on a marginal level each variable contributes equally to energy consumption. Energy intensity, on the other hand, had the coefficient with the greatest magnitude. This means that, on a marginal basis, energy intensity has the largest impact on total primary energy consumption.

DISCUSSION

Standard economic theory finds that energy prices will increase as the supply of natural resources used to produce energy, such as oil and natural gas, decreases. This supply–demand relationship creates a price signal that will encourage investment into the research and development of new energy-efficient technologies that will reduce energy consumption. In the long run, these technologies will lead to lower energy intensities for households and firms (Velthuijsen and Worrell, 2002). The end result will be an improvement in environmental quality, through a reduction in the consumption of natural resources, with a minimal effect on the economy (Foster, 2000). National energy policies around the world have been formed on the premise that price signals will create technological innovation that will reduce energy consumption. However, are these policies focusing on new energy-efficient technologies to reduce energy consumption, especially those policies aimed at specific sectors of the economy, the solution, as many stakeholders believe?

This chapter has presented an analysis that provides empirical evidence that the Jevons Paradox may exist in many countries. These results are significant given the concern about global warming, the increasing number of energy blackouts, rapidly increasing energy costs, the peak oil fears and the ever increasing demand for energy. The information provided in this chapter illustrates that technological improvements may not be the universal remedy that policymakers have been counting on. A variety of regions and countries were presented in this chapter to illustrate how widespread the Jevons Paradox may be. The countries included in the case studies were both economically and geographically diverse. The case studies include a developed country with a mediocre record on environmental conservation, a developed region with a strong environmental record, developing countries on the verge of 'developed' status, and a developing country actively promoting environmentally sensitive energy policies. The results strongly suggest that energy-efficient technological improvements as the solution for the world's energy and environmental problems will not work. Rather, energy-efficient technology improvements are counter-productive, promoting energy consumption.

Yet energy efficiency improvements continue to be promoted as a panacea. Consider, for example, some of the policies of one of the case studies presented in the chapter, the case of Europe. Europe has long been considered the leader in promoting a reduction in energy consumption and environmental pollution. Consider some of the energy policies that the European Union has adopted. For example, it has developed a policy that all members must aim at achieving energy savings of 9 per cent by 2012 through energy efficiency measures and has created initiatives like the ManagEnergy Initiative and the Sustainable Energy Europe Campaign 2005–2008 (European Commission, 2007). Furthermore, new EU members are expected to reduce their energy intensity and energy consumption levels to those consistent with other member states. However, on the basis of the results presented earlier in this chapter indicating that the Jevons Paradox may exist in Europe, one can conclude that policies promoting energy efficiency itself will probably not reduce energy consumption. While the examples of policies presented here are recent initiatives, a long list of prior regulations promoted and implemented in European countries could be presented as well. If one of the leaders in promoting and regulating energy consumption and environmental pollution is likely to experience the Jevons Paradox, then what are the conditions in other countries? And the other three case studies presented illustrate just how widespread the Jevons Paradox is.

However, this is not to say that energy-efficient technologies should not be promoted or sought after in very specific cases. If individual energy consumption behaviours are *significantly altered* to reduce consumption and this behaviour is unwavering, then energy-efficient technologies can further reduce energy consumption. In this case, technological improvements should be viewed as a potential complement to other energy and environmental policies. However, without a significant change in consumer behaviour, as has been shown throughout this book, energy-efficient technologies are very likely to lead to increased energy consumption.

The micro details as to why the results presented in this chapter suggest that the Jevons Paradox exists for the different case studies are not discussed here because such an examination of the evolution of the structure of each of the individual economies and societies would require a deeper investigation. Furthermore, obtaining such information is difficult as little is known, particularly on household consumption of energy. However, the research presented here is important nonetheless. This research will help identify the effects of national and regional energy policies so that energy strategies can be evaluated properly.

REFERENCES

Allan, G., Hanley, N., McGregor, P., Swales, K. and Turner, K. (2007) 'The impact of increased efficiency in the industrial use of energy: A computable general equilibrium analysis for the United Kingdom', *Energy Economics*, vol 29, no 4, pp779–798

Berkhout, P. H. G., Muskens, J. C. and Velthuijsen, J. W. (2000) 'Defining the rebound effect', *Energy Policy*, vol 28, pp425–432

Binswanger, M. (2001) 'Technological progress and sustainable development: What about the rebound effect?', *Ecological Economics*, vol 36, no 1, pp119–132

Birol, F. and Keppler, J. H. (2000) 'Prices, technology development and the rebound effect', *Energy Policy*, vol 28, pp457–469

Brannlund, R., Ghalwash, T., and Nordstrom, J. (2007) 'Increased energy efficiency and the rebound effect: Effects on consumption and emissions', *Energy Economics*, vol 29, no 1, pp1–17

de Haan, P., Mueller, M. G. and Peters, A. (2006) 'Does the hybrid Toyota Prius lead to rebound effects? Analysis of size and numbers of cars previously owned by Swiss Prius buyers', *Ecological Economics*, vol 58, no 3, pp592–605

Ehrlich, P. R. and Holdren, J. P. (1971) 'Impact of population growth', *Science*, vol 171, pp1212–1217

European Commission. (2007) http://ec.europa.eu/energy/demand/index_en.htm

Foster, J. B. (2000) 'Capitalism's environmental crisis – Is technology the answer?' *Monthly Review*, vol 52

Georgescu-Roegen, N. (1975) 'Energy and economic myths', *Southern Economic Journal*, vol XLI, pp347–381

Giampietro, M. and Mayumi, K. (2005) 'Jevons' paradox and complex adaptive systems: Exploring the epistemological conundrum when modeling the evolution of hierarchical systems', unpublished manuscript

Greene, W. H. (2000) *Econometric Analysis*, 4th edition, Prentice Hall, Upper Saddle River, NJ

Greening, L. A., Greene, D. L. and Difiglio, C. (2000) 'Energy efficiency and consumption – the rebound effect – a survey', *Energy Policy*, vol 28, pp389–401

Grepperud, S. and Rasmussen, I. (2004). 'A general equilibrium assessment of rebound effects', *Energy Economics*, vol 26, no 2, pp261–282

Gujarati, D. (2003) *Basic Econometrics*, 4th edition, McGraw Hill, New York

Haas, R. and Biermayr, P. (2000) 'The rebound effect for space heating: Empirical evidence from Austria', *Energy Policy*, vol 28, pp403–410

Haas, R. and Schipper, L. (1998) 'Residential energy demand in OECD-countries and the role of irreversible efficiency improvements', *Energy Economics*, vol 20, no 4, pp421–442

Hicks, A. (1994) 'Introduction to pooling', in T. Janoski and A. Hicks (eds) *The Comparative Political Economy of the Welfare State*, Cambridge University Press, Cambridge, UK

Howarth, R. B., Haddad, B. M. and Paton, B. (2000) 'The economics of energy efficiency: Insights from voluntary participation programs', *Energy Policy*, vol 28, pp477–486

Jaccard, M. and Bataille, C. (2000) 'Estimating future elasticities of substitution for the rebound effect', *Energy Policy*, vol 28, pp451–455

Jevons, W. S. (1865) *The Coal Question: An Inquiry Concerning the Progress of the Nation and the Probable Exhaustion of our Coal-Mines*, 3rd edition revised by A. W. Flux (1965), Augustus M. Kelley, New York

Kennedy, P. (2003) *A Guide to Econometrics*, MIT Press, Cambridge, MA

Khazzoom, J. D. (1980) 'Economic implications of mandated efficiency in standards for household appliances', *Energy Journal*, vol 1, pp21–40

Laitner, J. A. (2000) 'Energy efficiency: Rebounding to a sound analytical perspective', *Energy Policy*, vol 28, pp471–475

Murray, M. P. (2006) *Econometrics: A Modern Introduction*, Pearson Addison-Wesley, Boston, MA

Otto, V., Loschel, A. and Reilly, J. (2006) 'Directed technical change and climate policy', Fondazione Eni Enrico Mattei, Working Paper no 81, June

Polimeni, J. M. (2007) 'Jevons' paradox and the implications for Europe', *International Business and Economics Research Journal*, forthcoming

Polimeni, J. M. and Polimeni, R. I. (2006) 'Jevons' paradox and the myth of technological liberation', *Ecological Complexity*, vol 3, no 4, pp344–353

Polimeni, J. M. and Polimeni, R. I. (2007a) 'Energy consumption in transitional economies (Part I): Jevons' paradox for Romania, Bulgaria, Hungary and Poland', *Romanian Journal of Economic Forecasting*, forthcoming

Polimeni, R. I. and Polimeni, J. M. (2007b) 'Energy consumption in transitional economies (Part II): Multi-scale integrated analysis of societal metabolism and Jevons' paradox for Romania, Bulgaria, Hungary and Poland', *Romanian Journal of Economic Forecasting*, forthcoming

Roy, J. (2000) The rebound effect: Some empirical evidence from India', *Energy Policy*, vol 28, pp433–438

Saunders, H. D. (2000) 'A view from the macro side: Rebound, backfire and Khazzoom-Brookes', *Energy Policy*, vol 28, pp439–449

Schipper, L. and Grubb, M. (2000) 'On the rebound? Feedbacks between energy intensities and energy uses in IEA countries', *Energy Policy*, vol 28, pp367–388

Small, K. and Van Dender, K. (2006) 'Fuel efficiency and motor vehicle travel: The declining rebound effect', University of California-Irvine, Department of Economics, Working Paper no 05-06-03

Velthuijsen, J. W. and Worrell, E. (2002) 'The economics of energy', in J. C. J. M. van den Bergh (ed) *Handbook of Environmental and Resource Economics*, Edward Elgar, Cheltenham, UK, pp177–194

Wirl, F. (1997) *The Economics of Conservation Programs*, Kluwer Academic, Boston, MA

Zein-Elabdin, E. O. (1997) 'Improved stoves in Sub-Saharan Africa: The case of the Sudan', *Energy Economics*, vol 19, no 4, pp465–475

5
Conclusion

The Jevons Paradox can be associated with the concept of 'Malthusian instability', an expression coined by Layzer (1988). Malthusian instability refers to metabolic systems which are able to reproduce themselves (all living systems). It indicates that when operating in favourable conditions, they will unavoidably surpass the carrying capacity of their environment. This entails that the resulting process of natural selection will determine a constant evolutionary stress on metabolic systems. That is, there is a natural tendency of living systems to 'get in trouble', and this is the mechanism that enhances their ability to adapt and become something else. At the level of the individual metabolic system, this Mathusian instability is made possible by the existence of activities that in energy terms provide a positive return. In technical jargon we can say that these activities are generating a positive feedback or an autocatalytic loop. For example, if you invest 10MJ to perform an activity and you get 100MJ in return from this activity, and then if you re-invest the 90MJ of energy profit in doing more of the same activity, then you will get a total energy return of 900MJ. It is well known that a positive feedback like this one cannot go on unchecked for a very long period of time. We can recall here the story of Zhu Yuan-Chang's chessboard: if you put one grain of rice on the first square, two on the second, four on the third, and keeping doubling the number each square, there would be an astronomical number of grains of rice required for one position even before the 64th square is reached. This metaphor says it all. Hypercycles, or positive autocatalytic loops, when operating without a coupled process of control (and damping), do not survive for long; they just blow up (Ulanowicz, 1986). On the other hand, positive autocatalytic loops are required to provide the required supply of energy for those activities that are useful, but that implies a net loss of energy.

When researching the sustainability of the energetic metabolism of socioeconomic systems, Georgescu-Roegen introduced an analogous concept to define a typology of technology that can lead to Malthusian instability: the concept of 'Promethean technology' (or the viable energy technology). A technology is viable, just like a viable biological species, if and only if this technology can reproduce itself with a surplus of energy after being set up by the technology that is now in use (Georgescu-Roegen, 1978).

According to this definition, the feasibility of a technology is not sufficient for defining its viability. For example, a technology for the direct use of solar energy, which implies a deficit in the overall balance of energy over its life-cycle assessment (since other types of energy coming from outside the direct use of solar technology are required for its operation), would be feasible, but not viable. According to Georgescu-Roegen, in human history we have had only three Promethean technologies:

1   husbandry (agriculture);
2   the mastery of fire; and
3   the steam engine (or more generally the mastery of internal combustion engines), coupled to fossil energy.

These technologies share a common explosive characteristic: 'with just the spark of a match we can set on fire a whole forest. This property, although not as violent, characterizes the other two Promethean recipes' (Georgescu-Roegen, 1992). Land is the special fuel for agriculture. Fossil fuels are the special fuels for modern industry. As illustrated in Figure 5.1, due to the autocatalytic nature of Promethean technology, humans were able to get into the Malthusian instability trap quickly by depleting the special stocks of 'fuels' associated with these different technologies. In particular, the explosive characteristic of the petroleum-based metabolism of modern society, due to the abundant supply of high-quality oil in the past 50 years and the continuous supply of technological efficiency improvements, has been boosting the phenomena associated with the Jevons Paradox worldwide, as empirically shown in Chapter 4.

The Jevons Paradox is an issue that is little known outside of a few academic circles. However, the topic has taken on a new importance in this era of high energy prices, increased environmental awareness and concern over peak oil. At present, oil is the source of energy for nearly all the products that we consume. However, if the twilight of oil, vividly described by M. Simmons (2005), is really approaching in Saudi Arabia and the rest of the Middle East, oil producers will no longer be able to supply as much as the world will need, and we should start considering an alternative energy scenario to the conventional petroleum-based one.

Looking at Figure 5.1, one would expect that a quick inversion of the trends of energy consumption should take place soon, since exponential growth in the pace of consumption of resources cannot take place for a long period of time within a finite planet. Unfortunately, looking at the expected energy demand associated with economic growth (Figure 5.2), things appear to get worse in the future. In particular, developing countries in Asia are projected to have an annual growth rate of 5.4 per cent from 2004 to 2030 (Ito, 2007). According to Luft (2007), 58 per cent of China's oil imports come from the Middle East now and this share will grow to 70 per cent by 2015. China's concern for its growing dependence on oil imports has led to its active involvement in exploration and

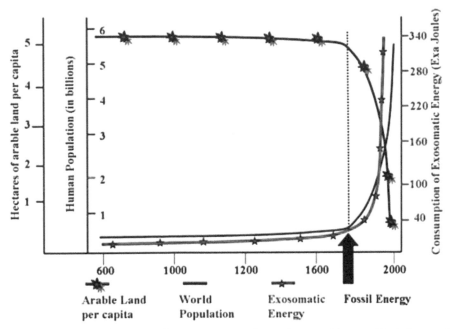

Figure 5.1   *The major discontinuity in the evolution of humankind due to the massive use of Promethean technologies*

production in places like Kazakhstan, Russia, Venezuela, Sudan, West Africa, Iran, Saudi Arabia and Canada. But China is not the only actor thirsty for oil in Asia; other countries, including India, are projected to be major contributors to the world's energy demand. In fact, India and China are estimated to account for approximately 70 per cent of the energy consumption in Asia over this 30-year time period (Ito, 2007). These projections should be a concern for many people.

The progressive reduction of oil reserves will force the world to turn to coal, natural gas and unconventional sources of oil such as heavy oil, shale oil, oil sands and tar sands. However, the supply of these energy sources is also limited, and the quality of many of them is much lower than that of current sources of oil. Nevertheless, while the world awaits a technological solution to the energy crisis, these alternative fossil energy sources will be consumed at an increased pace, creating an enormous amount of damage to the environment. Those alternative energy sources that do not rely on fossil energy are impractical at the moment because of their low energy return on investment (EROI) when considering their whole life-cycle assessment. In energy analysis EROI is the ratio between the quantity of energy delivered to society by an energy system and the quantity of energy used directly and indirectly in the delivery process. This index has been introduced and used in quantitative energy analysis (Cleveland et al, 1984; Hall et al, 1986; Cleveland, 1992). For example, the low EROI of nuclear power is due to the costs of mining and enriching the required fuel, building new plants and

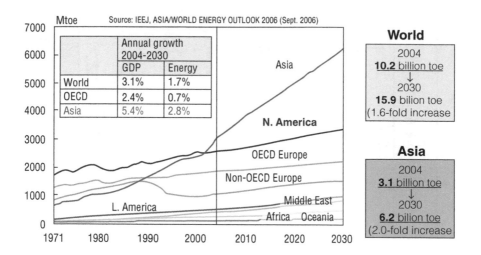

In 2030, primary energy demand of Asia achieves twice as much as current level, reflecting on highly economic growth. 3.1 billion toe(2004) → 6.2 billion toe (2030)

Figure 5.2   *World primary energy demand by region*

decommissioning old plants, plus handling and storing (practically for ever) nuclear wastes. The same problem of too low an EROI exists for biofuels and photovoltaic technologies. This is to say, for the moment many of the proposed technological alternatives to fossil energy are not viable technologies; they are only currently feasible because they are heavily dependent on fossil fuels. Yet traditional economic theory tells us not to worry, the increased price of these energy sources as their supplies diminish will lead to new technological sources of energy.

The information that has been provided in this book, from the historical overview to the theoretical and empirical implications, has illustrated that the Jevons Paradox is a very real threat to world energy security, as well as to the environment. Then, one may rightly ask, what is the solution? Should we stop looking for energy efficiency since this is a step in the wrong direction?

It is not our intention in this book to condemn technology or to suggest the 'right' way to go in relation to human sustainability based on scientific analysis. Rather, our purpose is to show that technology is not the cure-all that many people believe it to be. Technology can expand the option space for humans and by providing this service it represents an important complement to the energy-environment solution. For example, technological energy-efficient improvements such as hybrid cars may represent a crucial component of a different trajectory of evolution of the metabolism of households in developed countries. However, the adoption of hybrid cars per se will not solve the problem of the unsustainability of modern lifestyles if adopted by a world population of

9 billion people. If the energy-environment situation is to improve, consumers will need to change their behaviour patterns by including concern for the environment among the priorities determining their choices. In the same way, policymakers and other important stakeholders will need to accept the challenge implied by 'the tragedy of change' (as discussed in Chapter 3) which is typical of evolving systems. Humans have to accept losing something in order to be able to retain something else.

The need to deal with the tragedy of change, which is unavoidably associated with evolution, is essential in order to be able to deal with the Jevons Paradox. The big problem with the tragedy of change is that choices related to sustainability are choices that require reflexivity – the willingness to change yourself in order to be able to co-evolve with other humans and the environment. Unfortunately, choices based on reflexivity are not welcome in our modern society, since they imply dealing with the need for changing and re-discussing our own identity while dealing with moral issues. For this reason, humans try first to go for choices that, rather than being based on reflexivity, are based on externalization. The neoclassical economic paradigm is a good example of this: within this paradigm what is relevant is the willingness to pay to develop silver bullets and policies capable of preserving our own identity for as long as possible. But this paradigm leads to a dangerous form of denial of the obvious fact that humans have a responsibility over the choices that are made. According to the neoclassical economic paradigm, the process of becoming of humankind should be driven by the market and technological progress. By accepting this paradigm we can just keep doing what we are doing without thinking or reflecting on the consequences of our choices.

We believe that reflexivity is required to deal with the Jevons Paradox and is the only way to handle the issue of sustainability. Therefore, we hope that policymakers, the scientists giving them advice and other powerful stakeholders will take heed of the Jevons Paradox. In fact, the Jevons Paradox will always be with us, no matter what new energy sources and silver bullets we come up with in the future, especially when we discover another Promethean technology.

ACKNOWLEDGEMENT

The authors would like to thank Dr Kokichi Ito, Managing Director of the Institute of Energy Economics in Japan, for giving us permission to reproduce Figure 5.2. We are grateful for his kindness and support.

REFERENCES

Cleveland, C. J. (1992) 'Energy quality and energy surplus in the extraction of fossil fuels in the US', *Ecological Economics*, vol 6, pp139–162

Cleveland, C. J., Costanza, R., Hall, C. A. S. and Kaufmann, R. (1984) 'Energy and the US economy: A biophysical perspective', *Science*, vol 225, pp890–897

Georgescu-Roegen, N. (1978) 'Technology assessment: The case of the direct use of solar energy', *Atlantic Economic Journal*, vol 6, pp15–21

Georgescu-Roegen, N. (1992) 'Nicholas Georgescu-Roegen about himself', in M. Szenberg (ed) *The Life Philosophies of Eminent Economists*, CUP, New York

Hall, C. A. S., Cleveland, C. J. and Kaufman, R. (1986) *Energy and Resource Quality*, John Wiley and Sons, New York.

Ito, K. (2007) 'Setting goals and action plan for energy efficiency improvement' presented at the EAS Energy Efficiency and Conservation Conference, Tokyo, 18 June

Layzer, D. (1988) 'Growth of order in the universe', in B. H. Weber, D. J. Depew and J. D. Smith (eds) *Entropy, Information and Evolution*, MIT Press, Cambridge, MA.

Luft, G. (2007) 'Fueling the dragon: China's race into the oil market', *www.iags.org/china.htm*

Simmons, M. R. (2005) *Twilight in the Desert: The Coming Saudi Oil Shock and the World Economy*, John Wiley and Sons, Hoboken, NJ

Ulanowicz, R. E. (1986) *Growth and Development: Ecosystem Phenomenology*, Springer-Verlag, New York

Index